Going for Gold:

Pursuing and Assuming the Job of Fire Chief

Ronny J. Coleman

Delmar Publishers

an International Thomson Publishing company I(T)P®

Albany • Bonn • Boston • Cincinnati • Detroit • London • Madrid
Melbourne • Mexico City • New York • Pacific Grove • Paris • San Francisco
Singapore • Tokyo • Toronto • Washington

NOTICE TO THE READER

Publisher does not warrant or guarantee any of the products described herein or perform any independent analysis in connection with any of the product information contained herein. Publisher does not assume, and expressly disclaims, any obligation to obtain and include information other than that provided to it by the manufacturer.

The reader is expressly warned to consider and adopt all safety precautions that might be indicated by the activities herein and to avoid all potential hazards. By following the instructions contained herein, the reader willingly assumes all risks in connection with such instructions.

The publisher makes no representation or warranties of any kind, including but not limited to, the warranties of fitness for particular purpose or merchantability, nor are any such representations implied with respect to the material set forth herein, and the publisher takes no responsibility with respect to such material. The publisher shall not be liable for any special, consequential, or exemplary damages resulting, in whole or part, from the readers' use of, or reliance upon, this material.

Delmar Staff:
Publisher: Alar Elken
Acquisitions Editor: Mark Huth
Project Editor: Barbara Diaz
Production Manager: Mary Ellen Black
Art and Design Coordinator: Michele Canfield
Editorial Assistant: Dawn Daugherty

Online Services
Delmar Online To access a wide variety of Delmar products and services on the World Wide Web, point your browser to: http://www.delmar.com or email: info@delmar.com

COPYRIGHT © 1999
By Delmar Publishers
an International Thomson Publishing company I(T)P®

A service of I(T)P®

The ITP logo is a trademark under license
Printed in Canada

For more information contact:

Delmar Publishers
3 Columbia Circle, Box 15015
Albany, New York 12212-5015

International Thomson Publishing Europe
Berkshire House
168-173 High Holborn
London, WC1V 7AA
United Kingdom

Nelson ITP, Australia
102 Dodds Street
South Melbourne,
Victoria, 3205 Australia

Nelson Canada
1120 Birchmont Road
Scarborough, Ontario
M1K 5G4, Canada

International Thomson Publishing France
Tour Maine-Montparnasse
33 Avenue du Maine
75755 Paris Cedex 15, France

International Thomson Editores
Seneca 53
Colonia Polanco
11560 Mexico D. F. Mexico

International Thomson Publishing GmbH
Königswinterer Strasße 418
53227 Bonn
Germany

International Thomson Publishing Asia
60 Albert Street
#15-01 Albert Complex
Singapore 189969

International Thomson Publishing Japan
Hirakawa-cho Kyowa Building, 3F
2-2-1 Hirakawa-cho, Chiyoda-ku,
Tokyo 102, Japan

ITE Spain/Paraninfo
Calle Magallanes, 25
28015-Madrid, Espana

2 3 4 5 6 7 8 9 10 XXX 03 02 01 00

Library of Congress Cataloging-in-Publication Data

Coleman, Ronny J.
    Going for gold : pursuing and assuming the job of fire chief /
Ronny J. Coleman
        p.   cm.
    ISBN 0-7668-0868-8 (alk. paper)
        1. Fire chiefs.   2. Fire extinction—Vocational guidance.
    I. Title.
TH9119.C6496   1998
363.37'092—d321

98-38802
CIP

DEDICATION

This book is dedicated to three categories of fire chiefs. The first category includes all the chiefs who have ever served their communities. We need to remember them. The second category contains all the chiefs in service today in fire departments everywhere. We need to respect them. The third category represents all those chiefs who have yet to raise their right hands and swear an oath of office. We need to provide them with the wisdom, the wit, and the guidance to be ready when their time comes. To all three: Semper Vigilans.

As my career has evolved, there are three specific chiefs I want to recognize in this book: John Marshall, the chief who first promoted me to a gold badge, Chief Keith Klinger, who was an inspiration to me in my early days as a chief officer, and Fred Kline, Fire Commissioner for both Los Angeles City and County, who has served as a lifelong friend and advisor.

EPIGRAPH

I have no ambition in this world but one, and that is to be a fireman. The position may, in the eyes of some, appear to be a lowly one; but we who know the work which a fireman has to do believe that his is a noble calling. There is an adage which says that, "Nothing can be destroyed except by fire." We strive to preserve from destruction the wealth of the world, which is the product of the industry of men, necessary for the comfort of both the rich and the poor. We are defenders from fires of the art which has beautified the world, the product of the genius of men and the means of refinement of mankind. But, above all, our proudest endeavor is to save lives of men—the work of God Himself. Under the impulse of such thoughts, the nobility of the occupation thrills us and stimulates us to deeds of daring, even at the supreme sacrifice. Such considerations may not strike the average mind, but they are sufficient to fill to the limit our ambition in life and to make us serve the general purpose of human society.

—Edward Croker, Fire Chief,
City of New York

CONTENTS

INTRODUCTION

Being a fire chief in our society is a real honor. Do you really want to be one?

Those who have been selected to serve in that capacity are following in the footsteps of many generations of community leaders. The legacy goes back, in this country, for about three hundred and fifty years. On an international basis, it goes back even further, reaching to the days of the Roman Empire. Many of the founding fathers of the political system in the United States, such as George Washington and Benjamin Franklin, served in leadership roles in fire companies long before these companies were called fire departments. Succeeding generations have taken us through periods of change that were both traumatic and, at the same time, exciting. All of that experience was necessary to create the fire service of today.

In my time of service, covering nearly four decades, I have seen and experienced many things, suffered at times for taking unpopular positions, and felt the euphoria of accomplishment. All of these experiences have afforded me the opportunity to make many observations about what it takes to succeed as a fire chief. These observations are my opinion. I do not have scientific surveys to back them up. Instead, I have years of talking to some of the best and worst fire chiefs in my state, this country, and other nations as well. My observations are not necessarily facts; they are merely statements, but they are supported by the consequences of what I have observed.

What this book reflects is my attempt to provide an overview of some of the principles and practices that seem to be associated with *successful* chiefs. There may be people who have violated every type of behavior I describe in this book, and yet they have succeeded. There are always many ways to get a job done. I cannot quarrel with variation. In fact, I encourage it. However, I will say that I have used most of these techniques myself. They have not only worked but have also given me a great deal of personal satisfaction in the process.

The book is divided into two parts because I believe pursuing and assuming a fire chief's career has two different phases. These phases can be loosely defined as wanting and getting the job, then assuming command and setting your leadership and management style in place. Furthermore, I believe there are two other stages: keeping the job and leaving the job. Those are topics for another book. If people find themselves stuck in one of the first two phases for too long, they can get into trouble. Pursuing the job and never being successful creates frustration and disappointment. Just as importantly, however, assuming command is not a given. It has to be executed just like the pursuit or it can result in frustration and disappointment, too.

I have discussed most of what is in this text with my friends and peers over many years. Some of them are no longer in the fire service, some were never chiefs in the first place, some have passed away, and others are still in a fire chief's office somewhere, trying their best to achieve the most they can with the time they have left to serve.

When I was thinking about who I should recognize in the foreword of this book, I began to write down a list of the chiefs I have been fortunate enough to meet, work with, and observe. I found that I have thousands of names in my Rolodex that carry the title of Fire Chief. I chose to identify only three people who have served as my mentors over the years. Omitting all the rest does not mean I have not learned from them.

How were these fire chiefs selected? How successful have they been in achieving their goals and objectives? Has the job been as fulfilling and enjoyable as they had hoped it would be? If you ask those questions to a roomful of fire chiefs, you are going to hear a lot of different opinions about how it "really is." In some cases, they won the job by being highly competitive and beating out their peers. In other cases, they almost inherited the job. There are also a million stories that can be told about examination processes that were good, bad, or just downright stupid. There are many individuals currently holding the rank of fire chief who consider the process of selection to be arbitrary and subjective. Others are supportive of the various examination processes because they were required to go through a rigorous testing process and are proud of their accomplishments.

The concept of how fire chiefs are tested and selected has not been as widely researched as the testing and selection of police chiefs, so there is a lot of variety in how it is done in different states, cities, and districts. In my opinion, however, this subject may be irrelevant. The most important question to be asked and answered is whether a person is adequately prepared to do the job.

There are many who want the job of fire chief, but a much smaller number are emotionally equipped and technically competent enough to do the job well. How we select fire chiefs is really not important. What is very important is that the people who are selected be able to perform *after* they are picked. This is where job satisfaction begins. The process used to appoint the chief is not the process used to evaluate the chief's performance. While many pass the examination to get the job, some do not pass muster on the job. I say all of this in full awareness that there are about thirty-three thousand fire chiefs in this country, and possibly hundreds of thousands of them worldwide. Are they all doing the job of the chief in an environment that is euphoric? I would say that they are all trying to achieve that state, but few have the pleasure of coasting on the job.

I can say for sure that there are some chiefs who are better at their job than others. There are fire chiefs who look forward to going to work every day. They are enjoying the journey. Some leave behind them a trail of success stories in the form of day-to-day accomplishments that have made their communities safer because they have been there. Others leave behind a trail of broken promises, devastated individuals, and some loss of credibility for the fire service. How can we create more of the former and eliminate as many as possible of the latter?

Let us go back in history a few hundred years and take a look at another group that today is considered to be quite important in our society and see how this group evolved. Have you ever gone to a barbershop? You may have seen a red- and white-striped pole in front of the shop, and the pole may have been lighted or set up so it could rotate to attract attention. What is the significance of the red and white stripes in front of a barber's shop? The answer is found back in the 1700s.

At one time, when our society was relatively low in numbers, there were people who specialized in practicing medicine in our communities, taking care of small ailments by engaging in a process called bloodletting. It was fashionable at that time to blame illness upon the condition of the blood. If you had a fever, it was because your blood was too hot. If you were feeling lethargic, it was because your blood was too thick. Contemporary wisdom in those days said the best way to relieve the symptoms was to merely take a razor blade, make small incisions at the wrist and ankles, and bleed the person for a while.

Who would you guess got this job of bloodletting? Well, it was the same group of people who operated the barbershops. Many of these people served two purposes: there was not enough "medical practice" to keep them busy, and since everybody's hair grows, sooner or later everyone would need to get his hair trimmed. These individuals were simultaneously practicing medicine and hairstyling.

You should not have to guess too much to figure out what happened next. There were people who began to question the validity of bloodletting to cure ills. Some of the very same people who practiced those skills started doing research into other areas and establishing a minimum standard for those who were going to call themselves doctors. Schools were established. Groups of doctors got together and agreed on minimum standards of education, skill, and experience. Tests were created, and surgeons moved out of the barbershop and into the operating room.

The barber pole remains in front of the barbershop, but *surgeons* now operate in a totally different environment. They are still practicing medicine, but they have adopted very high standards that new surgeons must meet before being allowed to join the ranks of existing surgeons.

There was probably a point in time when some people had to make a choice about whether to remain barbers or rise to the new standards and become surgeons. Surely it did not happen overnight. As a matter of fact, it probably occurred over several decades. Some people stayed where they were, while some went forward.

The question of raising the bar for fire chiefs falls into this same category. We can continue to say that it is important for all of our subordinate personnel to be highly trained, even certified, in specific tasks and reject creating higher qualifications for ourselves at the fire chief level. It is conceivable, however, that not only will the certification processes below the rank of fire chief continue to thrive, but also that these specializations are going to demand more out of the fire chiefs of the future. What might happen if the fire chief does not have a strategy?

It is also conceivable that while we are failing to accept the need for creating some type of obligatory minimum standards for the fire chief's position, there are others, like city managers or elected officials, who may establish another set of standards by creating a new classification of individual who is looked upon as being more desirable to operate fire departments. The five-trumpet fire chief, as known in contemporary times, may be subordinate to a new class of individual who has accepted a higher standard that is more acceptable to society in general, such as a public safety director.

There will be people who disagree with that last statement. In the number of locations where we have seen public safety directors created in communities, however, that individual has seldom been the fire chief. That person has almost always been a person from the law enforcement community. One of the reasons why this may happen is that contemporary police chiefs have established educational criteria and are deliberately involving themselves in an arena that demands a much higher level of political acuteness and intellectual involvement than contemporary fire chiefs.

It is important that we increase the level of preparation for the fire chiefs of the future instead of worrying about the fire chiefs of the present. Now the question for those who want to be fire chiefs is "How should I be preparing to be a fire chief?" If we have all the other positions in place in the fire service, there is absolutely no logical reason why we cannot create a comprehensive career development process at the top of that pyramid for the fire chief. This position should be reserved for those who not only are willing to put their credentials on the line to prove that they are prepared for the fire chief's position but also actively seek to see their competency tested.

Recently, while studying career development processes in other nations, I discovered some interesting information. In the British Fire Service, the Fire Service College operates a course called the Brigade Commanders Course (BCC). It is specifically designed to prepare individuals to take over the commands of fire brigades in the United Kingdom. It is a sixteen-week curriculum that is intensive, demanding, and comprehensive.

One does not merely apply to take the BCC. Instead, one must compete *to become eligible to join* the course. Currently, the Fire Service College accepts nominations from brigades in the United Kingdom as well as other nations. When an individual's application is received, that individual is invited to participate in a three-day examination to determine *whether the individual is prepared* to enter the BCC. This simply means that not everyone who wants to go to the course gets to go. This also illustrates the fact that career preparation in the British Fire Service recognizes that there is a top rung on the ladder.

There are many differences between the American and the British fire service. The use of the preceding example does not mean that we need to emulate the United Kingdom's process in its entirety, but it is important to recognize that the process used for the Brigade Commanders Course requires that a person compete to get into the course. Just having the time to take the course is not the criteria. The course is not for existing chiefs. The BCC is for potential chiefs. By ensuring that a chief officer completes this course in the career ladder before he is appointed chief executive, the British Fire Service has done a great deal toward bringing about a unique combination of uniformity and standardization. This process, coupled with a curriculum that brings out the most competent and capable individuals, essentially means that the average chief officer is extremely good at what he is expected to do.

If we tried to apply the same process to the career development ladder in the United States, there probably would be a great deal of resistance. Our society is much more open. Yet, ask yourself this question: Does everybody who wants to go to medical school get to attend medical school? The answer is no. In order to perform the job of a medical professional, a person must compete for the process of entering school. Likewise, does every person who graduates from law school get to become an attorney? Again, the answer is no. The person must first pass the bar. Regardless of your grade point average in school, you must still prove to a panel of peers that you have a sufficient grasp of the legal profession to perform at a minimum level of standards.

So, we can choose to do a couple of things. One is to have our various training and education systems create a capstone course that is a fire chief's command course—one that puts together the best knowledge and information we can obtain on politics, interpersonal dynamics, finance, economics, legal proceedings, and so on—and limit the number of people who can take the course. That, however, runs contrary to American philosophical and political thought. The other possibility is to create a body of knowledge and allow anyone who is interested to access the information, while also developing an intensive scrutiny and evaluation process to ensure that an individual is totally competent before that individual is allowed to put the title after his name. The former model is that of the medical profession, while the latter is that of the legal profession.

Strictly from a practical point of view, the second model has much more appeal. It allows for options and alternatives to be available to individuals who wish to pursue the fire service career. It does, however, have one note of finality to it. Individuals can fail the test. This is the ultimate scrutiny of a profession. At the risk of oversimplification, it can be said that a particular task that can be done by anyone will never be called a profession. On the other hand, it is elitist to allow only a limited few to do it. If there are standards for a particular task and one can meet them, it is a profession.

The balance we are looking for in the fire service is to do everything we can to make sure that people who rise to the ranks of fire chief are capable of performing in a manner whereby they can lead their organizations, providing quality service to their communities, and being an effective leader in the community. This means, essentially, that every generation of fire chiefs should be a little more qualified than its predecessors.

Another element that my research has uncovered has to do with the fact that, in many countries, different type of fire chiefs are assigned to different types of fire departments. For example, in Germany, once a city reaches a certain population level, the criteria for becoming the fire chief become increasingly sophisticated. This particular model is not totally inconsistent with American practice, either. For example, have we not recognized the fact that there are doctors and then there are specialists? Have we not recognized the fact that, in the medical profession, there are levels such as general practitioner and surgeon? The answer, obviously, is yes.

Therefore, we have a potential in the fire service of identifying some specific criteria that match all fire chiefs in all environments, and then setting some increasingly sophisticated criteria for people who wish to be certified to operate sophisticated organizations.

I have not worked out all the details of what these ideas suggest, but this book focuses upon some of the important elements of *preparing* to become a fire chief before getting the job. It emphasizes the various types of *behaviors* that are required to respond to the demands of the job, not just the list of tasks in a job description. I suppose if we were back in that conversation in a room full of fire chiefs, there are a million questions that people would raise about how things can be done or should be done. The question is rhetorical, however, because we have not even taken the first step toward creating a uniform system for preparing fire chiefs to become fire chiefs.

If the medical profession had not taken this step in the 1700s, we probably would not be doing heart transplants, or have paramedics on our engine companies, today. Our society would probably be doing something totally different in the field of medicine. Likewise, a decision point for the fire service is going to occur sooner or later. Either we will take the initiative as the leaders of the fire service and go down the path of creating a more comprehensive preparation program, or there will be a parallel mechanism that will eventually allow someone else to take control of the destiny of the fire service. The very simple reason that this is going to occur is that the fire service is becoming an increasingly sophisticated occupation, and the consequences of failure are increasingly significant. The person who is leading a fire agency can no longer be there merely by reason of seniority or political influence.

I realize in writing these words that they are probably considered heresy by some, but I am sure that the person who took the first step and said it was time to stop bleeding people to cure headaches suffered criticism of a similar nature. I am proposing that *we* take the first step—that we establish standards for ourselves. We are the ones best able to put our hands on the bootstraps. We already have a process in place for creating standards for every other rank in the fire service. Career development guides are adequate and complete up to the rank of chief officer. It is time for us to act individually to become better-prepared to take over the job of fire chief.

We should not delude ourselves into believing that the fire chief of today is the model to which we are aspiring. As good as many contemporary fire chiefs are, they are not the future. The skills, knowledge, and behaviors that have been essential for survival in the past are changing. There are new skills, knowledge, and behaviors needed for the future. What we should be looking at is the model for the fire chief of the future—ten, maybe even twenty, years from now. What we can do is learn from the past and the present.

The process suggested in this textbook is aimed at individuals who may not even be in the ranks of chief officers yet. Some of the readers of this book may be coming to the peaks of their fire service careers at about the same time there is going to be a major exodus of senior fire officials from the fire service. Currently, each year we lose about ten to fifteen percent of the fire chiefs in the state of California alone. That is about a hundred fire chiefs a year. Nationally, the number is probably in the thousands.

If we begin to establish a career development process and to raise the level of expectations and the minimum skill requirements for people to achieve before they become chiefs, a natural winnowing process will occur. There will be some people who aspire to become fire chiefs but are simply not willing to work hard enough for it. There will be people who are amazingly intelligent but lack the commensurate human relations skills to do the job. There will also be people who have average knowledge, skills, and ability but are willing to make the deep-rooted commitment to develop themselves so that they are adequately prepared when given their gold badge.

Now that you have read this far, you may be asking yourself, "What next?" The reason I am expressing this concept in this book is that you, as the reader, are probably the type of person who is willing to make that commitment to try to make professionalism a reality in the fire service. I have had many conversations with fire chiefs regarding their feelings about the future. Almost all have expressed to me a sense of frustration, sometimes because of events that occurred just as they were about to end their own careers. Others have expressed doubts that the next generation is going to be able to handle the challenges. Then there are the optimists. I am one of those. We believe that almost every generation of fire chiefs has felt these same frustrations and fretted about its successors. History books in my collection are full of similar complaints from almost every generation of fire service leadership. The optimists among us are looking at the process of change as an improvement in the species; fire chiefs of the future *ought* to be better than the ones in the past. It is important for the survival of the entire profession.

I am asking you to engage in a process of rethinking how we prepare for and succeed as fire chiefs. Think about it more deeply than you would the process of just filing to take an examination. Intellectualize the concepts. I also hope you take this as an open invitation to be critical of the concepts contained in this book. Nothing worth having is ever achieved easily. I am hoping that you will sit down and go through this text with an eye on critical thinking. Identify every reason why these ideas will or will not work, especially for you. Then, I ask you to take the next step, which is to develop your own alternatives that will allow you to take control of your career. You can overcome the obstacles, no matter how high they seem to be at every stage.

This book is not about theory. It is about applications. Having edited the *Managing Fire Services* book for the International City and County Management Association (ICMA), and having served as a contributor to the last edition of the *Fire Chief's Handbook,* I am very familiar with the materials that have been offered for the development of chief officers. I have a different goal in this book. Although I refer to materials in other textbooks, the emphasis is on the nitty-gritty of competition, challenge, and change that a fire chief faces on a day-to-day basis. In some cases, the material is critical and irreverent of the role of the fire chief. In other cases, the book focuses on how to retain and improve upon traditional values of the job.

Many books on the topic of fire service career development seem to be focused on methodologies, technologies, and specifics that a person must learn in order to pass the written examinations. Many focus on dealing with the technical tasks of running a fire department, that is, prevention, operations, training, budgeting, personnel management, and so forth. This book does not refer to the technical aspects of the job except to highlight the chief's role in managing and leading those trying to accomplish technical functions.

Most general management books are devoid of specifics regarding the way in which a person actually applies leadership and decision-making concepts on a daily basis to a fire service agency. After writing several hundred columns for *Fire Chief* magazine on how a fire chief functions in the real world, and after listening to hundreds, perhaps thousands, of fire officers tell me their problems and how they have applied the information in some of my columns, I feel that I can provide a guidebook that will be very practical and highly personalized. I will let you be the judge of that after you have completed it.

—*Ronny J. Coleman*

SECTION I

Preparation

SOMEDAY THESE FIRE WAGONS WILL NOT NEED HORSES. THERE WILL BE A FIRE PLUG ON EACH CORNER, THERE WILL BE FIRE LAWS, AND THE FIRE CHIEF WILL HAVE TO BE A MANAGER... MARK MY WORD.

CHAPTER 1

When Were the Good Old Days?

A Historical Perspective

The objective of this chapter is to give a historical overview of how individuals have been tested and selected for the position of fire chief in the fire service. The chapter consists of an overview of processes starting from the late 1700s, when fire chiefs were first elected to their posts, up to contemporary times, when very sophisticated processes are employed to conduct executive searches. It includes an overview of types and styles of examinations.

THE GOOD OLD DAYS!

Do you remember the "good old days"? That expression has been used a lot in firehouses, usually by senior members of the organization when talking nostalgically about how much better things used to be before something happened in contemporary times.

Having lived and worked through several decades of the fire service, I have had the opportunity to listen to several different generations talk about their good old days. This, coupled with my interest in history, has given me an interesting perspective on what constitutes the "good old days."

Many of you probably remember a time in your career when things seemed to be perfect. However, how many of us recognize that period of time when it is actually occurring? There is a tendency for us to regard some events, as we are actually experiencing them, as difficult or less than enjoyable, and then to give them a whole new spin once we place them into our memories. As a result, the most memorable of our experiences are often traumatic when they occur but are pleasurable when they are recalled.

Another characteristic of nostalgia is that it is generational. There is a tendency when we are younger and more naive to be more open to new experiences and to the pleasures of those experiences; however, as we grow older and find ourselves becoming more confident, we begin to find ourselves more easily dissatisfied with conditions. This is true even when the conditions are better than those that preceded them.

Now what really gets interesting is when we combine a couple of generations who start comparing notes. Everyone has heard of the "generation gap," and it exists in the fire service like everywhere else. It is manifested when an older group of individuals has a different set of values and perspectives than its younger counterparts. Sometimes, it produces outright conflict. At other times, it is just a source of minor irritation.

What about those good old days? I started my career in 1960. There will be readers of this book who started their careers after that, because I am now one of the old guys. There are many individuals who are in the fire service today who were not even born at that time. On the other hand, I once attended a dinner where they announced one attendee's day of entry into the fire service; on that date, I was only ten months old.

When I first received compensation for being a firefighter, I worked a ninety-six-hour workweek and received the grand sum of $219 a month. I was hired on a Monday morning and driven to my first fire station without the benefit of any orientation or training. I was told to enter the fire station by the person who drove me there and to introduce myself to the crew on duty; the driver did not even bother to come in with me. I did not have a clue about my job duties or the degree of danger that was involved in the kind of fire fighting I was about to enter.

The agency I first started with was a wildland fire-fighting agency. Therefore, it was pretty much a seasonal operation. As a result, almost all of the people I worked with were not much older than me. The little knowledge they possessed was passed on to me through "on-the-job training."

In 1962, I left the wildland fire-fighting operations and entered municipal fire protection. Was it much better? Well, let us see. In the first place, back in those days the department did not issue any protective clothing. You had to purchase it yourself. I had to buy my first helmet, turnout jacket, pants, boots, and a complete uniform out of my own pocket. The price of that clothing equaled my entire first month's salary.

The first structural fire I went to occurred about fifteen seconds after I walked into the fire station on the first morning. As I entered the back of the station, a Klaxon horn went off, and a total stranger emerged from the back of the building. He glanced at me and my brand new protective clothing and asked, "Are you the new guy?" I muttered something in the affirmative. He said, "Jump on a rig, we're going to a fire." My initial response was, "What rig?" A pointing finger directed me to the back step of a piece of fire apparatus that was already running. I was standing alongside another person, whose name I did not even know. I was struggling to put on my turnout coat and jacket, hanging onto a chrome bar over the hose bed with one hand, as we left the apparatus bay and made a right turn. The person I was riding with was doing likewise. Shifting hands in mid-bounce, we both ended up with our coats on, but it was a miracle that we were not ejected into the street as we immediately made a sharp, left-hand turn. Were those days really the good old days?

Frankly, I can recall my first couple of days in the municipal fire service with such clarity that I can almost reproduce conversations that occurred between myself and others. It was a time of newness and vitality that has probably never been repeated in my personal experience since then. Yet, in retrospect, some of the things we did in those days were very dangerous and, in some cases, outright stupid. We did not realize that either of those words applied to us. We simply thought that it was the way things were done, and we were invulnerable. We must have been. After all, we never wore breathing apparatus because that was a sign of being a wimp.

Every day, we went out on the apparatus floor and cleaned the equipment. As soon as we had polished the floor to reflect an absolute mirror image, we would do crazy things such as wash and wax the tires. I recall that a captain had us use black shoe polish to shine the tires on the apparatus. When the apparatus was backed into the fire station, we literally got down on our hands and knees and wiped the treads of the apparatus so that there was no dust on the treads when the truck was sitting still. Do you think I'm kidding? I am not.

There were other equally bizarre behaviors on our part. I recall times when we had to get out of our work uniforms and get into a full class-A uniform, complete with tie, so that we would present a certain image to the public. A person in the driver's seat of an apparatus going anywhere never moved unless everyone was wearing a Class-A hat.

And what fires we had! We had fantastic fires—fiberglass factories would burn like college bonfires, and fruit-packing sheds went up with a great deal of regularity. Back then, if we spilled a flammable or hazardous material on the surface of the road, it was not a big deal. We merely called it a wash-down and made sure that, by the time we hosed off the asphalt, there was no residue on the surface. We cared very little about what happened to the material once it was in the storm-drain system. On "first-aid" calls, everybody got oxygen, no matter what kind of injury he had received. We even called it "scoop and run."

We had to fight very hard for all of the health benefits, salaries, and working conditions "given" to us in those days. The city refused to give a retirement system to the fire department during my first ten years. When we had meetings to discuss labor issues, they were almost always well attended, sometimes with as high as eighty to ninety percent of the off-duty force in attendance.

What does all of this have to do with being a fire chief today? In the first place, a lot of talk about the good old days as if those were the days when everything was so much better than what we have today is nonsense. Some claim that firefighters had it easier in the old days, that fire chiefs had it easier in the old days. The fact is that this is not true.

The contemporary fire service has evolved into an occupation that is more sophisticated than it was in earlier times, but it has never been, and never will be, an easy job. It is true that the policies, procedures, and processes we use today are almost all based upon experiences of the past. Such things as recruit orientation, in-service training and education, code enforcement processes, and protective clothing, including the daily uniforms we wear and the protective breathing apparatus we utilize, all evolved from solving the problems of the past.

We have made great strides. The compensation we receive for being combat firefighters and the benefits that society has offered to us are significantly greater than they previously were. Yet, there are often expressions of dissatisfaction with the way things are. Older generations keep wishing we could go back to the good old days. Younger generations seem to feel that what we have today can be taken for granted, and according to many of them, it is totally inadequate.

Like many other controversies, the answer lies not in the extremes but rather in the proper balancing of perspectives. No matter where we are in our fire service careers, we may look at the past and remember our experiences fondly, but that was then, and this is now. We should remember that each generation of firefighters paid a price for the accumulation of those experiences. It is all right to be nostalgic about the experiences we have accumulated in our lifetimes, but it is inappropriate to imply that somehow things were much better then than they are today.

Younger generations who are collecting their experiences at the front end of their career also need to maintain a proper perspective. There should be more consideration of the legacy and respect for previous generations, because they paid the price for what we have today. I recall a training officers' meeting where a younger officer kept referring to his seniors as "the old dummies." His lack of respect for their experiences and contributions finally raised the ire of a fire captain by the name of John Whelan. Captain Whelan, Los Angeles County Fire Department, decided he could take it no longer. He strode to the microphone and gave the group a tongue-lashing, basically as follows: "We might be the old dummies to you, but it is because we are the only ones who survived long enough for you to be able to criticize today. We paid a price to learn the lessons the fire service now takes for granted." He stated that he would rather be an old dummy who has made a difference than a young dummy who has never made a contribution.

Each generation, as it enters the fire service, has a great deal of potential. Conflict between the generations over what constitutes the state of perfection is counterproductive. The more important aspect of generational contribution can be based on respect of the past and pursuit of the future.

There are many who feel the fire service has lost certain dimensions that once made it a more enjoyable profession. Much of the camaraderie of the fire service has disappeared. Much of the respect that was once demonstrated by the community for the fire service has dissipated. Much of the loyalty and respect that was demonstrated between leaders and followers in the fire service has become mired in bitter conflict.

As I look back over my thirty-eight years in the fire service, I can say that if I had to do it all over again, I would not change a thing. I would still enter the fire service. I would still engage in the types of activities that have contributed to my experienced profile, and I would still advocate that there are others who should follow in the same path.

In the final analysis, the good old days were not necessarily good; they were just the old days.

SO YOU WANT TO BE FIRE CHIEF, DO YOU?

To put this into the context of aspiring to a fire chief's job, we might ask the following questions: Who was the first fire chief? How was he selected for the job? What did he have to do that is different than what we have to do today? What was it like in the good old days?

Unfortunately, history does not record the very first person who called himself a fire chief. But history does record the fact that someone was almost always in charge of the forces that had the task of combating fire. Blackstone's *History of the British Fire Service* notes that the first fire brigade was created by the Roman Empire to protect its cities. This group called the "vigilēs," was commanded by a military officer, a "tribunal." This officer was a person of significant rank within the military who was granted this rank by demonstrating competency under combat conditions. Interestingly, the tribunals that were assigned to run the vigilēs were of "equestrian" rank, which meant they were on horseback, as opposed to performing like infantry. They wore white tunics.

With the fall of the Roman Empire, there were very few records of fire service organizations in the cities and towns; therefore, there was little need for a person to be in charge. When the cities and towns began to emerge from the Middle Ages, the concept of using volunteer fire forces began to reemerge. However, there were really no requirements for the person who ran these organizations, either.

In fact, during the Great Fire of 1666 in London, the king himself got involved in the fire-fighting operation. He was in charge because he was the king, not because he knew anything about fire protection. Probably the first examples of organizational structures that began to look like fire departments emerged in the early 1700s. These organizations took on different looks in different parts of the world. In mainland Europe, the military took the lead. For example, in France, the fire service was the responsibility of the army and navy. In England, fire services evolved along the lines of the private sector because of the influence of fire insurance. There was a pretty sophisticated fire service in China at that time, but it was more a function of the royalty than the community.

If the history books are correct, the very first laws that were put on the books in the small communities that made up our original colonies were fire laws. Therefore, someone to enforce them must have been a part of the early community leadership in this country. Reportedly, the two things that created terror in the hearts of the early colonists were an attack by the native Americans and the prospect of a major fire, especially in winter.

Benjamin Franklin has often been described as the first American fire chief. He never took a test to become the fire chief. In the first place, nobody called him fire chief. In the second place, it would have been difficult to test for a position for which the job functions were not yet defined. Benjamin Franklin formed his first fire company in Philadelphia in about 1735. As you will recall, this country was undergoing a lot of change at that time, mostly driven by the desire for personal freedom. In the earliest eras of the American volunteer fire service, the process of selecting the top individual was much more democratic. As a matter of fact, the more I read about the early days of the American volunteer fire service, the more I am convinced that the whole concept of election of our leaders may have contributed in the revolution against the tyranny of England. A country that felt it could elect its leaders to protect it from disaster may have felt it had a right to elect others in power. Democracy may have had its genesis in firehouses.

Election was the favorite method of selecting the head of the department in those days. The chief was picked by a popular vote of the membership. Granted, there was little distinction between the person who was selected as the head person and the individuals who were part of the volunteer fire force. The level of fire protection was very unsophisticated. Apparently, there was nothing wrong with using the election process versus any other degree of evaluation, because the volunteers had to believe in the person that they were following. This process established at least one of the elements of the American fire service, and that was the sense that the chief usually came from one of the ranks. A brief review of history will demonstrate that, while we were splitting from the European continent on political grounds, there was still a lot of exchange of fire protection technology and methodology between Europe and the Colonies. In Europe, they created insurance companies. We did, too. In Europe, there were advances in fire pump technology; we adopted them, too.

Therefore, it should come as no surprise that many of the leaders in our political setting were also community leaders. We talk a lot about Benjamin Franklin, but Thomas Jefferson purchased a fire truck and placed it in his community, too. George Washington was a member of the Friendship Fire Company in Alexandria, Virginia. The list can go on and on.

What is important about this era is that the people who led fire organizations were leaders. They were almost always elected to that position. They were given the top job because they were regarded as individuals who commanded respect. As the society that bred the need for fire services changed, the fire services themselves changed. In Europe, the first paid fire officials began to appear around 1825. Among the first was a Scottish fire officer by the name of James Braidwood. Braidwood was made the Firemaster of the Edinburgh fire brigade, not just because he was popular with the volunteers, but because he was extremely knowledgeable in construction and mechanical trades. The process of selecting almost all of the first paid fire officials in Europe was based upon competency in these two fields.

In this country, we did not have any full-time fire departments until after 1850. For about one hundred and fifty years of our existence as a fire service in this country, the primary attribute for the fire chief was popularity. We had some wonderful individuals who acquired that level of recognition and support. The length of this chapter precludes an exhaustive coverage of this era, but a few paragraphs are essential. For example, Jim Gulick, a volunteer in the New York fire brigades, was so popular that he once led a major shutdown of the brigades merely by taking off his fire helmet on the steps of city hall. There was confrontation between the department and some political leaders over a policy. Gulick stated he would resign if it was not resolved in a certain manner. The story goes that, when his helmet hit the ground, a second or so later there was a crashing roar as thousands of helmets joined his. The issue was reconsidered by the community leaders.

I am not suggesting that leading a work stoppage is a good idea for a fire chief, but I am suggesting that a person who commanded that much respect and loyalty from his personnel must have been some kind of leader. A person who served as fire chief in the volunteer era had to have at least one of two qualities. He had to command respect, or he had to have some political hold on the group. The former was desirable; the latter could be a problem. For example, Boss Tweed of the infamous Tammany Hall was an elected foreman of a fire company who represents all that can be bad about having leaders who are elected. He took the political force created by his fire company and elected a group of individuals who were so corrupt they are still used as examples of political irresponsibility.

With the advent of paid fire departments in both Europe and this country, however, the popularity requirement began to evolve into a performance requirement. Once again, there was an evolution to the process.

We have to go to Europe to find the person who was probably the first fire chief to be selected based on pure competency. The first Firemaster of the Edinburgh fire department was the aforementioned James Braidwood.

Braidwood, who came from a family of carpenters, masons, and building construction tradesmen, had made a name for himself as an effective fire-fighter in the Scottish capital. He was selected to become the city's first paid fire chief, although chiefs were called Firemasters in the United Kingdom. Unfortunately, Braidwood, who also developed the first training concepts used by the fire service, suffered the fate of many firefighters—he was killed in the line of duty at a warehouse fire in London in later years.

Braidwood had migrated from Edinburgh to become the first Firemaster of London by the mid-1800s. He was succeeded by a gentleman by the name of Sir Eyre Massey Shaw. Shaw was actually a military officer with a high degree of personal charm and political savvy. He was selected to become the fire chief on the basis of all three of those characteristics because he was extremely competent. He set the ground rules for many years to come.

Shaw was given the post based upon his reputation for getting the job done. He kept it because he was able to demonstrate that his discipline and mechanical knowledge were good for the department. In this country, the very first paid chiefs were almost always selected from the ranks of the senior volunteers, many of whom also had military experience. Many evolved from being the elected chiefs to being the paid chiefs because they were capable of performing, but many did not. Popularity is a fickle criterion. Many individuals who acquired the job because of their social skills soon found that their knowledge of budgeting, mechanical, and construction trades was even more important. Some quit in disgust, some were fired in disgust, and some evolved into competent chiefs.

The concept of selecting fire chiefs using criteria other than popularity really began to emerge with the creation of paid fire departments. The paid fire service was spawned by a riot of volunteer firefighters in Cincinnati, Ohio, in 1851. The city government was so disturbed by the rowdiness of the volunteers that it was decided to replace them with a full-time force and, at the same time, to adopt fire steamer technology. The blow to the fire service was severe. In the first place, it took the concept of the fire chief out of the hands of the fire department membership and put it into the hands of the elected officials. Secondly, the implementation of a new technology, that is, fire steamers, meant a tremendous decrease in the need for staffing of large volunteer fire departments. This may well have been the first example of the so-called "downsizing" that we have experienced in the fire service.

It was not that the volunteer fire chiefs had not developed very large constituencies and power bases. Some of the more popular fire chiefs in the late 1800s were selected by their peers but were highly respected by their elected officials. The concept of picking the top firefighter as the fire chief did not necessarily evaporate when the selection process moved from election to selection. Generally speaking, the individuals who emerged as the first paid fire chiefs were individuals who had made reputations for themselves as very effective fire ground operators. "Gentleman" Jim Gulick, as well as many of the volunteers in the large metropolitan areas, was well known not only by the elected officials but also by the population in general.

If you look at pictures of these early fire chiefs, they almost all have the same serious demeanor. Few were really young people. A lot of them wore beards and mustaches. In most of the pictures, they look stern, even forbidding. However, the written narratives of the time clearly paint a picture of chiefs as being active, curious, aggressive individuals who were excited about their job of creating the American fire service. Once again, the history books tell us of chiefs like John Damrell, who in 1873 called an assemblage of fire chiefs into his city of Baltimore and suggested that they form an association so that they could exchange ideas more efficiently. That was the birthplace of the National Association of Fire Engineers (NAFE), an organization that was later called the International Association of Fire Chiefs (IAFC). What is interesting is the use of the term *fire engineer.* That was what the European chiefs called themselves at the time. The term *fire engineering* has a definite connotation to it. It is a discipline. Early fire chiefs apparently saw their role as having a very large component of fire engineering knowledge.

If we examine the era from 1850 to the early 1900s, we see that there were a lot of paid fire chiefs who were selected in various communities throughout the United States, but there was no uniformity in how they were selected. After the Civil War, almost all of the larger cities had paid fire forces. Going back to the concepts of popularity and competency, there is no doubt that most of the individuals who were selected as fire chiefs were able to demonstrate their skills and abilities to the satisfaction of the elected officials, or they would not have stayed in their positions. There are marked cases where individuals were "unselected" as fire chiefs and went their unheralded ways, for a variety of reasons.

Of course, this was the same era of the adoption of the steam fire pumper and the invention of both the electric fire alarm system and the automatic fire sprinkler system. These new technologies may have influenced how fire chiefs saw themselves at the time.

The development of the concept of actually holding an examination for a fire chief paralleled the use of the same process in industrialized America for selecting their senior executives in the field of organizational and personnel administration. More and more after the turn of the century, it became clear that individuals who wished to succeed as organizational leaders must meet some kind of minimum threshold to demonstrate vocational competency. Originally, however, most of the examination processes were relatively limited in scope. In some cases, the selection process consisted of a written examination based upon the department's policies, practices, and procedures. Later, the process incorporated the use of oral interview techniques. More often than not, the oral interviews consisted of neighboring fire chiefs who were invited by a community to sit as a board to determine the qualifications of someone who would soon become one of their peers.

What did evolve with this approach was the idea that a person who was to be the chief had to have some specific knowledge. Personal charisma started to take a backseat but did not leave entirely. It is noticeable that many departments in the United States retained the concept of a volunteer fire chief throughout this era. In fact, to this day, there are thousands of volunteer departments still active in the country. Many still elect their fire chiefs and are very satisfied with the results.

What also evolved with the examination approach was the requirement that paid fire chiefs possess certain types of knowledge, although a comprehensive description of that knowledge was lacking at the time. In general, during the late 1800s and even up until the 1930s, the chief was most commonly tested on his knowledge of the particular city in which he was serving and information about the technology used by that community. Notwithstanding the increase in examination processes, there was still a tendency, in the early 1900s and on up to the 1950s, to place a high premium on seniority in the business. More often than not, the person who migrated to the top of a fire department was someone who had spent a lot of time in the ranks. This was especially true in larger organizations. In order to overcome the tradition and bureaucracy within many fire departments, it was virtually impossible for a fire chief to be brought in from the outside. For approximately half a century, almost all of the individuals who became paid fire chiefs started at the bottom as entry-level firefighters and worked their way up through the ranks one at a time.

Professional mobility, or moving from city to city, was also restricted a great deal by the benefits, or lack thereof, of pension and pay. Initially, almost all of the retirement systems that came into existence were funded by local communities, and there was no ability to transfer from one community to another. Compensation and benefits for things like long-term health care and the aftereffects of injuries on the job were virtually nonexistent for the first half of this century. For almost fifty years, it was virtually impossible to rise to the fire chief's position without having extensive tenure in the same department.

Chief Ralph Scott of the Los Angeles Fire Department, when he was president of the IAFC, proposed that there should be a recognized body of knowledge. He obtained approval for defining this body of knowledge through the federal government, and the first set of professional standards were written for this country's firefighters. What this process did was shift the focus from how a particular city performed a specific task to how a specific task was performed by the fire service in general. In short, the creation of professional standards opened the door to the idea that there are acceptable national standards on how to do things.

In the aftermath of this change, the fire service began to change in many subtle ways. Almost all of the national organizations that represent the various aspects of the fire discipline began to create, publish, and distribute documents that could be used to define a person's expertise in the fire service. The result was that examination processes began to expand to include these documents. A person who wanted to move up the ladder now had to study for it by reading documents published by outside agencies like the National Fire Protection Association (NFPA) and the International Fire Service Training Association (IFSTA). Competition took on a different flavor. While experience still counted, education and knowledge began to command more consideration in preparing for the chief's job in the more sophisticated cities. Accompanying this increased emphasis on education and knowledge was an increase in the use of written examinations all the way down to the entry level of the fire service. In Shaw's British fire service, the one requirement was that an entry-level firefighter be strong and agile. In addition to the strength and agility of the candidates of this era was the additional requirement that they possess certain knowledge and skills. These examination processes also included the creation of other screening devices like oral examinations and physical examinations. The bar was being raised on getting in the fire service, so the bar was being raised on getting the top job, too.

For about another fifty years, the selection process for fire chief was pretty much left up to local governments to create and implement. The first reform that indicated a change here was the beginning of fire chief candidate recruitment outside of the department. I do not know which department first broke the tradition, but I clearly recall, during my early days as a firefighter, hearing others talk about the opening up of examinations to the outside. People were still scandalized by this in the 1960s.

The 1950s, 1960s, and 1970s were the next era of change in the selection of fire chiefs. The end of the Korean War saw one of the first major changes in the quest for the development of fire officers. With the end of the war came the G.I. Bill of Rights and an opportunity for veterans to go home and go to school. The idea of attending classes and competing for scores and examinations soon raised the level of competition within a peer group that contained command-level officers during this era. Notably, at the same time, the examination process for fire chief began to focus more and more on technical skills rather than time in grade. Granted, most of the individuals who were able to pass these extensive written examinations were initially individuals with a great deal of experience. However, over time, the grasp of technical knowledge began to diffuse itself into other ranks in the fire service.

During this time period, there was a significant increase in the number of educational opportunities for firefighters of all ranks. College degrees became available, and the educational component of the selection process increased. A review of job flyers from this time period reveals that departments were slowly starting to increase their use of educational criteria as a requirement to even take an examination. This created a dilemma of sorts. With a college education comes an extensive examination process, so the idea of using technical examinations as a screening device was not all that useful. Technical examinations leveled the playing field for candidates, but did not always evaluate other important criteria for success.

Concurrent with the increase in technological competence of chief officers was the significant increase in the number of fire agencies that went from volunteer to paid. Although there is no scientific data to back up this statement, it is pretty clear that between 1940 and 1960 a significant number of fire departments that had previously been volunteer began to develop full-time paid staffs. This is especially true in areas that bordered some of the metropolitan areas impacted by the growth of defense spending. It was also not uncommon for top-level chief officers in some of the larger communities to migrate out and become the first full-time paid chiefs for some of the suburban departments that were evolving from previous volunteer organizations. On the one hand, this was a diffusion of the technical skills that the larger fire departments were able to spread around in a short period of time. The corollary to that, however, was that many of the departments opposed the "big city approach," and there was actually a lot of conflict during this period of time.

Through the 1960s and early into the 1980s, the way that individuals prepared to become fire chiefs was a combination of both experience and education. There was a concentrated effort to improve the level of professionalism in this country. This was also the time frame in which *America Burning*[1] was written and the era in which most professional standards were created and adopted by many of the states for entry-level firefighters. The

---

[1] *America Burning*, The Report of the National Commission on Fire Prevention and Control (Washington, D.C.: U.S. Government Printing Office, 1973).

net effect of these improvements was to continually raise the level of competitiveness for individuals who wished to be chiefs of fire agencies.

The solution that was offered by many people in charge of personnel selection was the use of a technique called the "assessment laboratory." The assessment laboratory started to become popular in the late 1970s and hit its stride initially in the early 1980s. This type of selection process was oriented around the idea that knowledge must be demonstrated in order for it to be properly evaluated. The primary criticism of paper-and-pencil tests and even oral interviews was the fact that they did not adequately assess the depth of the real person behind the job. The concept of the assessment laboratory came from the military in World War II. It originally was used by many of the military organizations to select their candidates for highly specialized jobs, especially organizations such as the Central Intelligence Agency (CIA).

The assessment laboratory was designed around the idea that you could simulate different kinds of work behaviors and that candidates would be able to provide you with some insight on their performance by doing what they needed to do in front of a group of "assessors." The concept had a lot of merit. These assessment laboratories were supposed to be simulations of the types of tasks that a fire officer faced on a daily basis. The "exercises" were things like in-baskets, staff meetings, disciplinary events, fire ground emergencies, and so on. The process used outside observers to score the performance of the candidates based on "dimensions," which were types of observable behaviors that test creators believed were indicative of the appropriate ways to deal with the simulated problems.

Assessment laboratories allowed, more than anything else, a reinstatement in the selection process of a dimension of personality and character. Charisma and style, in spite of the best intentions of the assessment laboratory designers, were allowed to come back into the discussions of who was the best candidate for the chief's job. And, as with other forms of testing, individuals began to study the test instead of studying for the job. There was a whole body of knowledge accumulated about how to take assessment laboratories so that you could score well. Unfortunately, these classes did not place a lot of emphasis on the candidates' integrity, and as a result, the assessment laboratory became more of a role-playing scenario for people who were good at role-playing instead of being good at being chief.

Another phenomenon that paralleled the assessment laboratory was the creation of a group of people who went out looking for candidates to take these tests. They were called "executive recruiting firms," or "headhunters." This concept was already in place for other industrial jobs as early as the 1930s. Its use increased with industrialization and the growth of certain occupations. It is a simple idea. Instead of going out and testing the universe to find the right person for the job, an organization finds a headhunter whose job consists of understanding the occupation and, at the same time, knowing about the talent pool that is available. The goal of the headhunter is to go out and select individuals so that the screening process has a higher possibility of success. This concept became a part of government with the increase of the city-manager form of government. Executive recruiters contributed to the selection process by narrowing down the number of candidates forwarded to departments that opted to use the concept.

It became obvious, with the use of assessment laboratories and then the addition of headhunters, that it was no longer a necessity to start at the bottom of an organization and move up until one was ultimately given the chance to become chief. Instead, the concept of professional mobility began to set in. Communities became more and more willing to accept candidates external to their organization. Candidates became more and more willing to exchange one set of benefits for another. The net result was that in the 1980s and early 1990s, there was a great deal of professional mobility. One of the problems with this professional mobility and the raising of the required level of skills, however, was that the job of fire chief was becoming less and less fun. At one time, the job of fire chief was looked upon as very honorable, and hardly anybody in the community would ever think of challenging the fire chief's authority. During the period of time that the fire service was changing, however, so was society. It was a period of time when the all-white, male dominated perspective of who was the candidate for chief was first challenged. Minority status and gender became an issue with the selection of candidates. In the past, fire chiefs tended to be only men, and few minorities were ever considered. Public policy on civil rights began to change at about this time. Many of the personal benefits of becoming the chief were eroding. Becoming the chief became less desirable for both financial and emotional reasons.

Among the reasons that made the job of fire chief less desirable was the enactment of federal legislation that made it very lucrative to remain at lower levels in the organization. Oftentimes, lower-level chief officers made more money than the chief of the department because of overtime. There also was a considerable increase in the potential of the fire chief to be held accountable for something that his department did, thereby increasing liability. In addition, with the idea of professional mobility came the potential of conflict between labor and management, primarily because an individual who came into the department from the outside often did not have a lot of the institutional memory of the organization and therefore often struck out on support for issues that were foreign and distasteful to the labor group.

Votes of "no confidence" became more prevalent. It became less and less desirable to become a fire chief at the same time that it became harder to become one.

That pretty well takes us up to today. Right now, there are about thirty-three thousand fire departments in this country. There are some that still elect their chief officers. There are some that still require that a person be a member of that department to be a candidate for fire chief. There are some departments that still require a pen-and-pencil "civil service" type of examination. Some use executive recruiting firms. Some use assessment laboratories. What is a candidate to do?

Well, this might seem trite, but first you need to decide whether you want to go for the job before you worry too much about how to get it. Are you serious about becoming the fire chief? If there were any kind of slogan for a fire chief, it might be: I didn't promise you a rose garden. The job of fire chief is a lot more complicated and a lot more comprehensive than it ever was in the past. The nonfinancial motivation to become a fire chief in today's environment is driven much more by internal forces than it is by external ones. Some might say that no one in his right mind who is a competent chief officer, working fire suppression duty work or twenty-four-hour duty shifts, would give that up to become the chief of the department. But then again, perhaps some would.

What appears to be occurring today is that fire chiefs are selecting themselves and then entering the competition in order to find out if they will be given the opportunity to fulfill their own wishes and desires. I come from the school that says that there is no such thing as being "tested" for fire chief. You are selected based upon the fact that you are what you think you are, and whoever is hiring you knows exactly what you are too. That is the basis for much of this book. A person can be extremely intelligent and technologically competent and still be a total disaster as a fire chief. Others may be extremely personable and have all the charisma in the world, but lack the ability to be decisive. The fire chief of today is a rare combination of all of the skills and abilities that are needed to run a fire department, and is most successful when those skills and abilities are linked with an organization that needs that combination at some point in time.

The next big question to be answered is: How do I get ready for the competition? The answer is simple. When in Rome, take the test the way the Romans do! Preparation on the part of the candidate requires that you perform the essential task of reconnoitering the process to see what you are up against. There is no other way. You are in the driver's seat, in one way. You can choose the process within which you want to compete. The absolute truth is that you cannot be successful in a selection process using only one strategy. There are different criteria for success in each process, and if you wish to achieve the top score, you must prepare specifically for each process.

No matter what process is utilized by an organization to find them, the individuals who enter it must be as prepared as they can be. You do not study to become a fire chief. You study to be able to do the job of a fire chief, and then you are selected from among your peers in accordance with a lot of dimensions other than technical competency. Granted, we do not elect our paid fire chiefs anymore, but a modern fire chief has to have the ability to provide leadership and management at the same time. This involves being recognized by the individuals in an organization as a person who can lead. While we recognize the fact that fire chiefs must have served some time in the ranks to be able to adequately relate to the people they are leading, it is not absolutely essential anymore that they spend most of their career there. The fire service is slowly sorting itself into three groups of people: officers, individuals who probably would rather remain as firefighters for their entire career, and those who want to be the chief someday.

What is important for the fire service to recognize is that seniority is not the leading criterion for making a person the best fire chief of the future. My personal philosophy is that if you do each and every job you are assigned as well as you possibly can, you will become a candidate for upward mobility. It is unlikely that a person graduating from recruit academy looks down at their watch, glances up to a calendar, and then says, "Well, in twenty-two years, I will be the chief of this department." None of us approaches our career that way. Yet, we must make a conscious decision on our part to prepare ourselves to be fire chiefs, if in fact that ends up being our goal in life. I am not entirely sure when it happens.

In talking this over with many of my friends, I have found that, many times, individuals did not know that they would be interested in the job until the opportunity was right in front of them. That has to be somewhat scary. On the other hand, there are individuals who started planning and maneuvering their way to become the top dog so early in their career that they failed to gain the depth and experience that is absolutely essential as they moved upward through their ranks. There is a balance here somewhere. If I were going to produce a set of guidelines, it would consist of the following statements:

A. Do everything you possibly can to make yourself as competent as you can in the job you are doing today.
B. Prepare yourself as quickly as you possibly can to take the next position in the rank structure when your confidence level allows.

C. Never accept any more responsibility than your emotions will allow you to control.

If you want to become a fire chief, now is the time to think about it. If you already are one, perhaps it is time to reflect upon it. If you are really not sure, the material in this book may help you to decide.

HOW CAN YOU CONTROL YOUR FUTURE?

Jules Verne, Isaac Asimov, Ray Bradbury, Alvin Toffler, and John Naisbitt are authors with a specialty. Through trend analysis, each of them has provided a vision of the future. Although their works are never listed as required reading for fire protection professionals, these authors have much to say about our profession and your chances of becoming a fire chief. Why? Because yesterday's science fiction often becomes today's scientific fact; today's seemingly isolated event often becomes tomorrow's trend. The future, including the future of fire protection, is beginning to emerge today. The fire protection delivery system is undergoing change, so the role of the fire chief must be changing at the same time.

Thinking about the future of fire protection is not always comfortable, especially after one studies how outsiders view the profession's future directions. In *Fahrenheit 451,* for example, Ray Bradbury's fictional firefighters are relegated to burning books. John Crutcher, vice chair of the United States Postal Commission, once stated at a congressional hearing that "no monopoly can be assumed to last unchanged forever, impervious to changes in technology, consumer taste, and the larger economy." Public fire protection is largely a monopoly and faces a similar assessment. The Libertarian Party's one-time presidential candidate David Bergland made a campaign promise in one election to continue public police forces, but to *close all municipal fire departments.*

The fire protection delivery system is undergoing change and may, in fact, have changed more than any other public service or branch of government. In some communities, fire protection master planning and the mandated use of smoke detectors and sprinkler systems, for example, have shifted some responsibility for fire and life safety from the public sector to the private sector. The shift in emphasis from fire suppression to fire prevention is more than theory in many communities. The acceptance of responsibilities in the areas of emergency medical services and hazardous materials has stretched our organizational capacity to respond. More and more fire agencies are trading different solutions to fire problems through the use of technology.

A variety of technological changes that offer widespread impact on fire protection are either already available off-the-shelf or will soon appear. Sophisticated electronics can now speed the report of a fire alarm from the dispatching center to the fire station within seconds. State-of-the-art sprinkler heads now activate within moments after a fire's ignition. Soon, sophisticated fire ground robots may perform tasks that are too dangerous for humans and, thus, free firefighters to do the fire suppression tasks that can be done only by people.

Fire protection technology is, in fact, rapidly moving from science fiction to scientific fact. I once wrote an article that described, as science fiction, the operations of a mythological engine company in the year 2033. Robots, computers, satellites, and avant-garde suppression chemistry were all discussed. I wasn't talking that far into the future. Most of the technology is around today, but isn't widely known or accepted.

Technology is often available long before it is accepted. We use technological innovation to improve our personal lives, yet fear that the same technology is dehumanizing when applied to our profession. There is terror in technology along with its promise of efficiency and a better quality of life.

Which technological changes will be accepted in our future? Who will control the decisions about the future of fire protection? The answers to these questions depend, at least in part, on how much the fire protection profession is willing to challenge its beliefs, how we visualize change, and who we permit to control fire protection's future.

Professionals have a view of themselves based on what members of the profession consider universal truths. These truths are sometimes called paradigms. According to Princeton University Professor Thomas Kuhn, major changes or breakthroughs occur only after someone has challenged a profession's paradigm. Someone says that the old truth is wrong and that there is a new truth. Those who dare to challenge the paradigm, if they are successful, change it to create a new truth.[2]

VISIONS OF THE FUTURE

Many of fire protection's basic assumptions have been challenged in recent years. It is no longer a basic truth, for example, that firefighters must be men or that residential sprinklers are a dream. Like many changes, the advent of female firefighters and the development of residential sprinklers began as

---

[2] Thomas Kuhn, *The Structure of Scientific Revolution* (Chicago: University of Chicago Press, 1996).

hope. Because different segments of the fire protection community have different hopes for the future, there is no single vision of fire protection's future. Instead, several visions of the future are emerging. One of the realities of becoming the fire chief is that you will have to create a vision of the future or fulfill someone else's vision.

Some people see the future as a slicker, cleaner, somehow better version of the status quo that fulfills their own wishes. Labor organizations, for example, hope for a future that requires significantly more firefighters than we have today. Equipment manufacturers hope for a future increase in sales as the private sector takes on more fire protection responsibility. Fire chiefs hope for a future that includes a funding panacea that will decrease their budget vulnerability.

Although these visions may be attractive, it is not enough merely to hope for a future. Like any profession, we must choose a future. The act of choosing often starts with a series of "what if" questions that challenge the current paradigm. What would future fire protection be like if a newly invented chemical or process could extinguish fire more effectively than water? What would happen if we had sprinklers in all of the world's high-rise buildings? What would happen if allied professionals, such as architects, had more technical knowledge about fire protection?

These questions are a necessary starting point for your preparation to become a fire chief in the future. In the language of change, the exclamation point is more powerful than the question mark. One implies a decision. The other implies a need to wait until there is an answer. Unless the fire protection community acts to control its own future, someone else will. For this reason, actions of future fire chiefs are essential to the future of our profession.

The best crystal ball is actually a mirror. In the words of the World Future Society, we need actions to show that we have "thought globally, but acted locally." If the fire chief of the future challenges the fire service's own paradigms through "what if" questions and then acts to make the changes that he believes in, maybe we can control the future of fire protection. Choosing to defend the status quo is an option, too, but it contains some dangerous vulnerabilities. History has already demonstrated to us that the fire service is evolving and that the successful fire chiefs are the ones who can lead that change.

The choices and changes you may have to make if you choose to be a fire chief will not be easy. While the very nature of fire protection demands some caution, the chief's job is to know how far to push the envelope of change. There is not yet agreement about what the future of fire protection could, or even should, be, but one thing is clear: the decisions and actions of fire chiefs today and tomorrow will shape fire protection of the future. Do you still want the job?

SELECTED READER ACTIVITIES

1. Conduct an interview with an existing fire chief to determine the chief's perspectives on the current job of being a fire chief.
2. Collect a dozen or more job flyers for the position of fire chief and review them in the context of your background and experience.
3. Review the historical files of your department to determine what processes and activities have been a part of selecting previous fire chiefs. Compare and contrast the differences between them over the last few decades. Compare and contrast them with the practices of neighboring or similar-sized fire agencies in the area.
4. Write down your vision of the future of the fire service.

CHAPTER 2

Getting into the Playoffs:

The Decision to Try for Chief

The objective of this chapter is to discuss the practical and political consequences once a person chooses to become a fire chief. This chapter expands on the changing role of the fire chief discussed in Chapter 1. It discusses the political expectations created by those who hire fire chiefs, and the changing role of fire chiefs in dealing with labor and management issues.

> *Who can doubt, if one knows all they are required to know to*
> *fight fire, that this is not a profession.*
> —*Sir Eyre Massey Shaw[1]*

GOING FOR THE GOLD

Ask doctors what they do for a living, and they are likely to answer, "Practice medicine." Likewise, attorneys practice law. Now for the $64,000 question: What do fire chiefs do for a living?

Well, if you have a sense of competition, you could respond in the same fashion. You could profess the same level of professionalism: "Fire chiefs practice fire protection." But the response begs the question: are we a profession?

Medical doctors and attorneys must meet very clearly defined standards before they can claim to be doctors or attorneys. The last chapter discussed what fire chiefs have had to do to be selected for the job. Do we really have a set of professional standards on a par with doctors and attorneys? Must we all meet the same standards before we get the gold badge? Are we a profession? Personally, I do not think so, but I am not embarrassed by that, either.

More than one hundred years ago, the chief of the London fire brigade, Sir Eyre Massey Shaw, was among the first to recognize that firefighters must be adequately trained and possess a certain level of discipline if they are to be given the responsibility of protecting lives and property. His text, *Fire Protection, Operations, Machinery and Discipline of the London Fire Brigade,* written in 1886, was among the first fire service textbooks to describe the body of knowledge required for a fire officer.

But what about now? Did we live up to Shaw's prediction? What real criteria is used today for the selection of professional fire officers, particularly fire chiefs? My answer is not going to be very popular. I do not think there are any criteria that are uniformly applied. The fact is, there are some standards applied to the testing, selection, and promotion of firefighters, fire officers, and even the lower-level chief officers. But with regard to the fire chief's job, the standards are not uniform.

---

[1] Sir Eyre Massey Shaw, *Fire Protection, Operations, Machinery and Discipline of the London Fire Brigade* (1886).

Granted, we have a set of professional standards that have been created by the NFPA for ranks in the fire service. Several states have created certification programs for different levels of fire officers. But nothing has stuck to the wall as far as establishing universal standards for selecting fire chiefs.

This chapter does not discuss what those standards should be, as that would be a daunting task all by itself. Instead, it focuses on the problem of individuals landing a fire chief's job when there are no real standards against which they can be measured. How do you prepare? In short, this chapter deals with the issue of getting a job when no one knows the rules of the game.

If there are no uniform standards, how do up-and-coming fire officers know when they are ready to take on the job of fire chief for the first time? How do they know when they could be competitive for a better job in another community? If there are no standards, how can the fire profession assess its successes and failures in advancing the state of the profession? These are heady questions, right?

But to get down to basics, if there are no recognized standards, then each of you must set your own. Setting your own standards is a somewhat different concept than meeting some external criteria. But by establishing and striving to meet your own standards, you are investing in your future; you are making yourself more competitive; you are making the position of fire chief more rewarding for yourself. And, finally, you can have a positive impact on the level of professionalism we are all seeking if you build a career ladder by accumulating the kinds of qualities you will need when you have the chief's job. So how do you go about setting your own standards? What should you be thinking about right now?

Rule 1. Adopt a set of ethics. Ethics are a set of rules, or a code of conduct, in which you believe. They are guidelines to your behavior. They are not just meaningless sentences or jingles and rhymes to frame and hang on the wall. Both the International Association of Fire Chiefs (IAFC) and the International Society of Fire Service Instructors (ISFSI) have developed sets of ethics. Adopt one of them, or develop your own.

It is important to adopt a code of conduct because your professional behavior is always going to be subject to evaluation. It will be evaluated by your peers, by your subordinates, and by your bosses. If your performance and behavior are based on a predictable set of standards, you will be able to develop a style of leadership that can be predicted. By urging predictability, I am not suggesting mediocre behavior but, rather, consistent behavior. When you live by a code of conduct, your honesty, integrity, work ethic, and lifestyle can be anticipated by the officials who select fire chiefs, since they will know who you are and what you stand for. Become a known quantity before you enter the selection process.

Rule 2. Set your educational achievement goals as high as possible. Look around you and determine what other people are doing to get ahead, then do more. Do not be lulled into believing that you are competitive just because you have the same type of educational profile as everyone else. Give yourself a competitive edge.

Consider the educational criteria for some of our peer groups, such as police chiefs, community development directors, city engineers, city attorneys, and city managers. Almost all of them have degrees of some sort, and those in the truly competitive jobs usually have graduate degrees.

That does not mean that you should run out and get just any degree, but it does mean that you should set your goals fairly high—well above the accepted norms for the education levels of your subordinates, and even above the norms for those of your peers.

Rule 3. Study the fire chief's job description, the city, and the organization before you aspire to a specific job. Do not be so enamored of the chief's gold badge that you aspire to the wrong job at the wrong time for the wrong reason. The real key to competitiveness is that the needs of the job should closely match your personal qualities and abilities. (More is said about this in Chapter 3.)

There is a common myth that in the competition for the chief's job, the "best man" wins. This is a myth because there are a lot of different kinds of "best" persons for different kinds of jobs. Moreover, some of the men are women. In actuality, the task of selecting a fire chief is just that— a selection process. The key to being a successful candidate is to compete in selection processes that are matched with your skills, abilities, and experiences. If you want to be a winner, enter races based on your particular type of handicapping.

This sounds like a commonsense statement, but the reality of it is often overlooked. I have discussed this issue with many unsuccessful candidates for promotional examinations who felt put down because they were rejected during the screening process, given poor scores in the oral boards or assessment centers, or rejected by the city manager upon final interview. Rejection does not mean failure. It just means that the requirements or dimensions established for the selection of candidates simply did not match a particular applicant's inventory. Did the person fail? No. The person was not selected because he was just not the right person for that position.

Study the city or district that you want to lead to determine its culture, values, and political ambiance. Study the organization of the city thoroughly to determine its fiscal policies, organizational philosophy, and the methods and mannerisms of the city manager and other department heads. Study the fire department to determine its strengths and weaknesses, its programs and directions, and its needs. Then compete or accept the jobs in those places where *your* characteristics match the characteristics of the job.

Rule 4. The best time to find a job is when you do not need one. The most successful preparation technique is to succeed in your current job. If you build your résumé through solid accomplishments, you will be amazed at how often you will be more competitive in promotional examinations. You may be recruited for the selection process, but not if you have a record of mediocre performance. The concept of head-hunting or executive recruitment is becoming more common in the fire service for select jobs. If you are recruited for a process through an executive search firm, your chance for success increases by quantum leaps.

If there is one thing that can damage your chances in the competitive process, it is the perception that you really need the job. It casts a pall of failure on the candidate. Desperation is never a good qualifier, especially if it actually results in a candidate getting the job. If members of an assessment laboratory or selection board detect that a candidate is desperately trying to escape something, they are likely to consider this a mark against the person. This certainly does not help a candidate's chances. But the real tragedy is when such a person is given the job and the person offering it knows that the candidate is desperate. The candidate is more likely to agree to do things that he would not have agreed to do otherwise. Many a fire agency has suffered because of this phenomenon. The greatest support mechanism you can have in the competitive process is job satisfaction in your present position.

Rule 5. Become and remain involved in professional organizational activities. If you are an aspiring fire chief, hang around with people who already are what you want to become. Where do you find them? The answer is, at professional conferences and meetings. For example, most states have an annual conference of their fire service. If you are in the New York fire service, go to the New York conference. Since most state organizations have similar conferences, the same rule applies no matter where you live. One of the greatest opportunities is at the International Association of Fire Chiefs annual conference where all states contribute to attendance. Or, if you are thinking about competing for a fire chief's job, you might want to become involved in organizations that contain city managers, such as the state's League of Cities or its equivalent.

Getting to know individuals in professional organizations helps in many different ways. You can always learn a lot, and you might identify a role model or locate a mentor. But nothing will happen if you do not get involved. By participating in such activities, you can develop a network of contacts that can increase your competitive edge and at the same time increase your visibility in the fire community.

So now you have five rules for becoming a more visible candidate for the top job in the fire station. Obviously there is a lot more to it than this, but if you follow those five rules fairly closely, you will become someone that others will fear to see on the list of competitors.

INVESTING FOR A RETURN IN YOURSELF

Following the rules provided in the preceding section is not enough by itself. You also have to remain current in the field to be competitive. You have to learn how to stay in touch with fads, trends, and patterns. For example, have you had enough of management clichés? We have been exhorted to "search for excellence" and to become "one-minute managers." We have gone through "future shock," and we have ridden "the third wave." It never seems to stop, and most of it is just a rehash of basic common sense.

It seems like every time we go to a conference or, worse yet, our superiors return from a conference, we are given a whole new set of buzzwords and clichés regarding our behavior and our future opportunities. In the field of public administration, it seems that there are a lot of people who lie awake nights to come up with new theories about how we should behave. Frequently some of these clichés are turned around and used as means of evaluating productivity and performance of leaders and managers. For example, Management By Objectives (MBO) evolved from a relatively simple concept into a massive paperwork monster in many organizations. Zero-based budgeting in many organizations meant "what the city giveth, the city shall taketh away."

So, is there a defense against this barrage of alphabet soup? How can we turn it around and make it an asset to the fire chief candidate? The simplest technique that can be used to protect the integrity of your own decision making is up-to-date information. Information is power. No matter what the latest dance craze says, the more you know about the way things really are, the more likely you are to be able to do things your own way. My suggestion on how to deal with this is so simplistic that it is probably overlooked by most of us on a daily basis. The solution is: invest in yourself!

What I am talking about here is for you to take a certain amount of your income and invest it into an information network that will allow you to know about things before they come through the bureaucracy from the top down. I am suggesting that you take a certain portion of your discretionary funds and obtain a combination of subscriptions and memberships that will feed you with intellectual ammunition.

One of the most common remarks from city managers, city councils, and other elected officials is that they are trying to make government work more like business. Almost all of the fancy terms and buzzwords that have come from the government bureaucracy recently have had their heyday in the field of private enterprise and have been experimented with by the federal government before they ever arrived on the desks of state and local government officials. Therefore, it makes sense that, if you have some kind of a hawk's eye view of what is coming through the communications network of the federal government and private enterprise, you may have advance warning of the things that are going to land on your desk someday.

The reason that it is important for us to evaluate our information network is that the average fire officer believes that most information is going to come through the professional organizations to which he belongs. In other words, we rely a great deal on professional journals such as the *International Association of Fire Chiefs, Fire Chief, Fire Engineering,* NFPA publications, the *International Society of Fire Service Instructors,* the *American Fire Journal,* and *Fire House.* Unfortunately, they do not always contain leading-edge material on management trends. You often have to find that information in other publications such as *Nations Cities* and *Governing* magazines. Granted, a lot of state and regional fire magazines basically repeat in a written form the latest "dance craze" that is going on in the fire service, but none of these professional publications really provide us with much advance warning. We should have access to all of them, but we should not expect them to be on the real leading edge of change.

When it comes to investing in yourself, all of the magazines listed in the preceding paragraph can be sent to your home address for a total investment of less than fifty cents per working day. If you are one of the many candidates who wait until the department subscribes to these publications, you need to change your thinking in this area. If you are familiar with the early warning system for the Strategic Air Command, you are probably familiar with the fact that the radar system we use to detect our enemies is way out on the front lines. It is not next door. The Strategic Air Command calls the system locations DEW lines, which stands for distant early warning lines. How about setting up a distant early warning device for yourself?

One of the tools that is available to you is a frequent buyer program in the general management textbook area. There are groups that are similar to the Book-of-the-Month Club whereby you sign up to receive management publications as they arrive on the market. For each month's selection, you are given the opportunity of either taking the book that has been selected for you or choosing another title. There is no obligation to buy a particular number of books as a member of most management book clubs. The price of the books and the available titles, however, usually generate a cost of anywhere from twelve to fifteen dollars on a monthly basis. Some of the books are rather basic, while others are right up there on the best-seller list. Most important, they are a reflection of a changing state of the art in management science. They are not written about fire; they are written about running organizations. Here is where the early warning line exists.

An expenditure of twenty-five to thirty-five dollars per month to obtain this kind of information is, in my opinion, inexpensive. If you were a medical doctor or an attorney, you would spend considerably greater sums than this just on monthly journals and subscriptions that would be required to keep your medical or law library up-to-date. Also, expenditures for books of this nature are tax-deductible; therefore, you get a double-edged asset. You have the information, and at the same time, you are able to write it off.

Another early warning mechanism is a business magazine entitled *Success*, which focuses on the success stories in the field of business and industry. It also provides frequent articles on trends and patterns in management methodology. Business magazines such as this one contain a wide variety of advertisements that point out available resources with which to keep abreast of changes in corporate and business philosophy. The current subscription cost of a good business magazine is approximately thirty-six dollars per year. This is about the price of one coffee break a week for an entire year. As mentioned earlier, it is a dual asset in that it provides you with early warning information and is also a tax deduction.

An excellent resource to keep you aware of what is changing in government is a series of publications put out by the International City and County Management Association (ICMA) entitled *Management Information Services Bulletins*. The MIS series focuses upon practices in communities that tend to be on the leading edge of change. For example, the concept of "quality circles" that was first generated in Japanese literature and in some of the European literature made its appearance in the MIS series long before it was talked about at fire service conferences. The business concept of strategic planning was addressed in the ICMA bulletins long before it hit the pages of fire publications. Almost all of the current and forthcoming management philosophies have been addressed by the ICMA in one fashion or another. One does not have to become a member of the ICMA to access this information, either.

If you recall from the movie *Patton*,[2] there is one scene in which General Patton was engaged in tank warfare in the African desert. He was reversing the trend of the German Panzer successes by using some of their own tactics against them. His line was, "Rommel, You . . . I read your book!" I am

---

[2] *Patton,* 20th Century Fox, 1970.

not suggesting that we are engaged in combat with city management; however, it is very important that we be on the same communications level as they are. By periodically reviewing the list of MIS bulletins that are available and actively pursuing the reading of the more pertinent of them, you can keep yourself in the proactive mode.

One last technique that I would propose for you to consider could be called the development of "professional curiosity." As you read professional publications such as a monthly fire magazine, you will frequently find a section that deals with book reviews or describes current literature. One of the ways that you can keep your in-basket full of information is to read this section critically. You can collect many free publications that are generated by other governmental and business agencies. I have a stock letter in the word processor that essentially refers to the issue of the professional publication in which I read the announcement, asks if I can receive a copy of the publication mentioned in that issue, and uses a generic acknowledgment thanking the organization for making this publication available.

When I was a young training officer, I got into quite a discussion one time about the cost of training. One of my fellow training officers told me, "If you think training is expensive, try *not* training someone and see how expensive *that* gets!" It is the same way with investing in your future. If you think magazine subscriptions or the periodic purchases of books to add to your management shelf are expensive, try not doing it and see how long you survive in the administrative and management jungle.

According to my records and receipts, it costs me less than five hundred dollars a year to access the information stream to which I have referred in this chapter That is little more than a dollar a day. You probably leave more than that in tips during coffee and lunch breaks.

The watchwords of the 1990s are very focused on performance, but in reality, it is often difficult to improve performance using just theory. When you become a fire chief, you will need resources—personnel and, most important, policies and procedures that work.

Fire chiefs' departments are often confronted with problems for which there are no instantaneous solutions. I am not talking about fire-fighting problems, because the fire service is pretty good at putting out fires. What I am talking about are all the other kinds of problems that a fire chief faces— for example, the problems of obtaining community support and legislative support, and the problems of solving difficulties that are not necessarily emergencies but are crises nonetheless. You need to be prepared to use three techniques in a fire chief's job, and in your current job as well. These are:

- problem clarification

- internal innovation
- external solution search.

Problem clarification is taking the time to define the real issues that are affecting the performance of your organization before you act. Unfortunately, the fire service tends to be oriented around solutions rather than problems. We tend to fix things by identifying a solution without clearly understanding the real problem. Problem clarification is a process whereby you sit down and take a look at an issue using Rudyard Kipling's five serving men: who, what, when, where, and why. It is important to identify what is causing the problem, when it is occurring, where it is occurring, and why it is occurring before deciding how to solve the problem. That is a skill you can develop long before you take the oath of office for the chief's job.

For example, let us look at a topic such as juvenile fire setting. There are many juvenile fire-setting programs that are available to the fire service. Will they work in your situation? There is no way of knowing the answer to that question unless you actually clarify your problem before you select a solution. In reviewing correspondence from a lot of departments, I have noticed that we tend to identify goals and objectives based on the perception of a problem, but seldom do we go to the depth of clarifying the problem before setting the goals and objectives.

Maybe this is the fault of how we have taught people about the goal-setting process. We have focused most of our attention on the establishment of the goal and the creation of the objectives, and we have not emphasized the fact that a trip of a thousand miles is a waste of time unless you know why you must go that way. Another example will allow us to try on this thought process.

Assume that a department is having problems with its budgetary situation. Money would seem to be the solution. On the surface, it may appear that the problem is that the department is just not getting enough money to do the things it wants to do. But that may not be the real problem. The problem may be that there is not enough money to service all of the issues the community is facing. Therefore, using the analytical process, the problem may be reshaped—the question of how to get more money may be restated as one of how the department can reconfigure itself so that it can live within the budgetary constraints.

The best way of starting the problem clarification process is to simply ask the question, what is the problem? Write it down. Then take each of the words in the sentence and ask yourself, is this truly an expression of the problem? For example, we might start by saying that we do not have enough money. The problem with that is: the city does not have enough money. The problem with that is: we cannot do everything we want to do. The problem with that is: we must decide what are the most important things to do and eliminate those that are unnecessary. The problem with that is: we must set priorities. The problem with that is: we will have to discontinue doing some things that may be traditional. The problem with that is: certain people with special interests will resist that change. The problem with that is: it will turn into political conflict. And so the process continues.

I once had a staff officer who used this technique quite effectively. Every time we proposed a particular solution to a problem, he always came back with "the problem with that is." Initially, it was frustrating to work with him because he seemed to raise more objections than he did solutions. Nonetheless, by allowing him the luxury of doing that, we were often able to work our way through the problem until we reached a point where he ran out of problem statements. We would make a statement, and he would just sit there with a smile on his face.

As one might expect, this process is intellectually intense. It requires that you not accept superficial answers. It requires that you continue to scratch away at the surface of something until you clearly understand the causes of a problem.

This opens the door for the second technique, *internal innovation.* You can find a million books on innovation and creativity at the local bookstore, and I am not about to suggest that this chapter is anything new about being innovative or creative. But I will emphasize one thing: innovation requires nurturing. It does not happen by accident. If there is not an environment in an organization that rewards people for seeking out alternatives, then it is unlikely that alternatives will be offered.

In terms of internal innovation, there are a couple of essential elements that must be in place to provide that nurturing. One is an open channel of communications where ideas can be floated through the organization without being cut off. There are a variety of means available to encourage such communication, including suggestion boxes, frequent meetings with personnel, open-door policies, and so on. The technique that is selected is not as important as the environment that is created.

The second element is what is referred to as "incubation." Incubation is nothing more than allowing ideas a sufficient amount of time to exist so that they can begin to grow. If an open channel of communications allows ideas to emerge, then incubation must exist as part of the process to allow them to mature. Some of the techniques that are available for incubation are such things as creating project reports, conducting staff analysis, and even encouraging experimentation and research projects. The period of incubation varies in accordance with the scope of the project. Probably the best way of expressing this is that big ideas take a long time to really come to fruition.

The concept that nothing is as strong as an idea whose time has come is a reflection of incubation. The way that this is institutionalized in a fire organization is to allow such things as pilot projects or demonstration projects to prove that innovations have merit and validity.

The third concept that we need to explore is that of the *external solution search*. How many times have we heard the expression that it makes no sense to reinvent the wheel? That is what we are talking about here. As I suggested earlier, belonging to professional associations, going to meetings and conferences, and attending training and education sessions are all valuable tools in leveraging outside solutions.

What we have identified in this discussion is a three-step process, but it does not really consist of three steps. It is more like three points on a compass. Coming up with innovative solutions to problems is not a linear function. It is cyclic. Problems do not go away; they merely change in shape and size. What this chapter is suggesting is that if we employ three basic techniques to try to keep our department on the leading edge while we are working at lower levels, we may be able to do it once we obtain the top job. These techniques consist of (1) making sure that we clarify our problems before we attempt to come up with solutions, (2) doing everything we can to encourage people internal to our department to come up with options and alternatives to solve problems, and (3) leveraging outside influence by literally borrowing other people's materials, making the necessary adaptations, and putting them to work as quickly as possible. These techniques create a cycle, since it is not uncommon to solve one problem and create another.

In researching my personal library, I found it interesting to note that solutions to problems freqeuently have a life cycle to them. For example, the current trend toward the public education concept in the fire service is really nothing new. As a matter of fact, if you go back to the early writings of the fire service in the 1870s and 1880s, there were several proponents at that time who thought we could solve America's fire problem if we could just change the attitude of our schoolchildren. It was innovative then; is it innovative now?

Do not forget, in the process of trying to come up with innovative solutions, to periodically go back and look in the history book. There is a lot of gold left in the tailings of previous administrations. All that is required is that you dust it off and put it into the context of a contemporary fire agency.

The next concept I would suggest that a potential chief consider is similar to one of the biblical parables. It is stated in the Bible that if you give someone a fish, you feed him for a day, but if you teach that person how to fish, he will feed himself forever. It is the same way with problem solving. If we, as fire officers, personally solve all the problems, we can possibly avoid the crisis of the day. But it is much more important that we teach our subordinates how to solve problems so that they will continue solving them long after we have moved on to better rewards. The concepts of problem definition, internal innovation, and external solution search are not tools strictly for you as a potential chief; they should be advocated as part of the toolbox for anybody who is in a position of management or leadership within any modern fire organization.

METAMORPHOSIS: INCUBATION OF AN OFFICER

One of the most amazing transformations that takes place in the world of nature is when the caterpillar becomes a butterfly. The process is miraculous and, at the same time, quite simple. The caterpillar serves time in a certain form, then spins a cocoon to go into an incubation period. When it reemerges, it is turned into a distinctly different type of organism. Just what goes on in that cocoon is a series of events that transforms one form of life into another. As soon as the butterfly emerges from the chrysalis, it is expected to behave in a different fashion. Instead of walking about on legs, it is now expected to fly.

The system expects the same thing of our individual fire officers once they move from the technical and task-oriented level to being responsible for the leadership of a fire department. In my career, I have promoted hundreds of individuals. I have always felt that they suffered somewhat from not being able to go through a cocoonlike stage in order to ensure a smooth and effortless transition. In the case of becoming the chief, the phenomenon is very pronounced.

There are not many opportunities for a cocoonlike stage to occur in the fire service at the lower levels. For example, when a person goes from apparatus operator to engine company officer, there are some new expectations, but they are fairly predictable. When an individual moves up to a position as a new battalion chief, supervising numerous fire captains, there is a higher level of expectation but no period of time to evolve. Then, of course, the ultimate jump from one lifestyle to another is when a person goes from being a subordinate to actually being the fire chief. It is even more complicated when a person becomes a fire chief of a city other than the one in which the person grew up.

We will probably never have the luxury of a cocoonlike phase for a fire officer. Many fire departments, however, have instituted policies, practices, and procedures that have a similar effect; they allow the individual to go through a transition period before being impacted by the new expectations.

The promotional process supposedly evaluates whether a person has potential for the next level. Yet, there is nothing inherent in the testing procedures to ensure success. In a lot of cases, the testing procedure criteria for measurement are not even job-related. So, what can be done to create a cocoon for the incipient fire chief?

There are at least three different techniques that I have seen used in fire departments to perform this function at the lower levels: officers' candidate schools, officers' orientation periods, and temporary assigned duties. Although few departments utilize all of these techniques, individual fire departments often utilize some of them. They are examples of the efforts made by fire officials to give new fire officers a perspective on their new roles once they have been pinned with the badge of responsibility.

Officers' candidate schools are exactly what they sound like. They actually consist of curricula that have been put together to teach fire officers what is expected of them once they achieve a new position. Not unlike their military counterparts, officers' candidate schools are often aimed at making sure that rank-and-file personnel (entry-level/experienced firefighters) are exposed to a new body of knowledge of what is expected at the fire officer level.

Notably, the majority of the officers' candidate schools are in organizations that have large numbers of potential fire officers. In some areas, when student population is not large enough to support an officers' candidate school, state fire training programs have developed curricula to accomplish similar objectives. Many state fire training programs have a fire officer curriculum that is based on NFPA standards. The purpose of this type of program is to provide potential fire officers with a new perspective on their roles.

Officers' orientation periods are slightly different from the candidate schools. Officers' orientation is a program that is used in organizations where an officer has already been promoted. In effect, it is a temporary assignment that places a fire officer candidate out of the normal organizational setting for a short period of time. The process results in the candidate being briefed by all of the other individuals in the hierarchy of the organization. For example, officers' orientation curricula might include such topics as familiarization with the fire prevention bureau, orientation to the department's training division, refamiliarization with the city's personnel practices and procedures, and perhaps time spent with the chief or chief officers of the department talking about mission and philosophy. The officers' orientation concept is based on the assumption that the officer has the basic skills and abilities to perform the routine details of the job but needs to have an overview of the organizational setting in order to become more effective.

Officers' orientation courses can be developed regardless of the size of the organization. Granted, there may be some expenditure of overtime and constant coverage, or there may be overtime expended to bring the incipient officer in on his day off to engage in the officers' orientation, but in either case, a one- to two-week period is usually sufficient to provide an officer with orientation of all the specializations with which they should be reasonably familiar.

The third technique, temporary assigned duty (TAD), is once again borrowed from the military context. Temporary assigned duty is often used in organizations to remove a fire officer from the fire company setting and to place them into either training divisions, fire prevention, public education units, or some other staff function for a period of time from thirty to ninety days. The TAD concept is similar to the orientation concept except that it is more task-oriented and is designed to give individuals exposure to a specialization so that they may carry certain skills and abilities back to their roles as fire company officers.

The type and size of organizations that exercise TAD assignment are directly related to their own budgetary practices and the philosophy of the fire chief. While this concept is probably not as widespread as the other two, it has been very successful in reorienting new officers to functions outside the realm of normal fire suppression and day-to-day operations.

There are other techniques to assist in the metamorphosis of an individual from a technical level to a supervisory or managerial level. It is not uncommon in many organizations for a person to be required to move from station to station or from shift to shift upon appointment. The theory behind this is to remove the individual from the close personal working relationships with individuals they served alongside prior to promotion. Some fire departments have even gone to the trouble of developing a reading list or task inventory, or both, for probationary fire officers to complete during their six- or twelve-month probation.

Unfortunately, the type of metamorphosis that we see from caterpillar to butterfly simply does not happen in the real world. One does not emerge a new and made-over person merely out of being promoted or from assuming new responsibilities. Yet, the theory is relatively sound: people must learn to change behaviors if they have changed responsibilities. To a large degree, the metamorphosis of a fire officer is entirely dependent upon personal commitment to this process.

In the context of this chapter, I would like to remind you that the caterpillar has to spin its own cocoon. It is not spun by other caterpillars, nor do the other butterflies come back to help. Individual responsibility marks the beginning of the metamorphosis process, and individual responsibility determines what a person turns into once he has moved from the more technical and task-oriented aspects of firefighting into the increasingly hierarchical levels of responsibility in fire management.

When I received a promotion from the position of firefighter to engineer, someone gave me a cartoon that contained a caricature of an individual with crossed eyes and buck teeth with the caption, "Yesterday I couldn't spell enjineer and today I are one." The more we work on developing processes to ease the transition from one level to another in the fire service, the more likely we are to be able to improve the productivity of organizations and increase the professionalism of the service as a whole. Emerging from the cocoon, even if it is a psychologically fabricated one, means becoming something that we never were before we were promoted. It also means that we can never go back to being what we are before we were given the new responsibility.

CLIMBING THE LADDER OF SUCCESS

Sometimes we just over complicate things. We tend to say that the more sophisticated something is, the more likely it is that only a very few people will be able to understand it. We develop whole vocabularies and languages that are mystical and ambiguous to keep people from understanding what is going on.

One of the best-kept secrets in the fire service is the concept of career development. Those who have been successful in acquiring the top jobs in the fire service, that is, fire chiefs, have somewhat preserved the mystique of it. People who are entering the fire service at the lower levels and looking up at that hierarchy often ask themselves, "How do you get from here to there?"

There is a concept called the Career Development Guide in many fire departments. I am not going to repeat the discussion of the need to identify specific courses of action that people need to follow in order to be ready for promotion. That is a lengthy subject in itself. I am going to focus on your own aspiration. This is a more subtle and less clearly defined behavior, but only you can decide to prepare for success in your career.

What I would like to do is compare aspiration in the fire service to something that we all learn in the basic recruit academy—climbing ladders. The type of ladder work we do in the fire service is very physical, and yet it serves as a metaphor for our ability to succeed in our lives. One thing you need to recognize right away, for example, is that some people do not have the strength to throw certain kinds of ladders. There are people who do not have the strength to perform as fire chiefs, either. Not everyone who enters the fire service has the strength to climb the ladder of success all the way to the top job. Sometimes there are individuals who get to the top rung, not because they have raised the ladder, but because the team with which they have worked has put them there. Aspiration includes your family and friends, as well as your desires.

The second thing you need to recognize is that ladders do not stand up of their own free will. They must lean against something. Historically we have used fire ladders to gain entry to buildings so we can perform rescue or ventilation to get above a fire. The analogy in career succession is that no career ladder is ever raised very high unless it is confronted with some challenge. The larger the challenge, the more likely it is that the person who must climb the ladder will need the strength we alluded to earlier.

Not unlike laddering a burning building, you need to recognize that career development often means putting yourself in harm's way. There are risks associated with climbing the ladder. Individuals who ignore those risks can suffer the consequences of failure and become casualties. Those who recognize the risks and take the necessary precautions to avoid damage are the individuals for whom we are looking. A person who arrives at the final rung battered, bewildered, and beaten will no longer be able to perform; a person needs to arrive there in good mental and physical condition in order to be able to do the most with that position. In a later chapter, we will talk about what they never tell you at the fire academy. Most of those things are items that will consume both your energy and your patience. You had better have both.

Now if we go to the other end of the ladder, recall that no ladder is safe unless it is firmly footed. The stability of our careers is not necessarily based on our competency as firefighters. Much of the stability of our careers centers around our family, friends, and those individuals who provide us with the mental and moral support in order to pursue the challenge. Failure to pay attention to this most fundamental safety feature of climbing actual ladders can often result in career ladders being destroyed. When the ladder falls to one side, the individuals who are fairly high up on the rungs may find themselves in imminent danger. Footing is a part of a ladder's stability. Pay as much attention to establishing trust and faith in your family as you do in learning how to answer questions on written and oral examinations and preparing for assessment laboratories.

Our next metaphor for life from ladders is the beams. The two beams we have on either side of a ladder are there to provide structural stability. They provide some equal spacing of the rungs we must climb, but they are always there to prevent collapse of the ladder. They are relatively rigid and unchanging; therefore, they are something we can hang onto as we proceed to climb the ladder. The two beams you must have for successful career planning are your *principles* and *values*. You need to be firmly grounded in these before you begin to climb. If you do not know what you stand for today, it is highly likely you will fall for something inappropriate tomorrow.

The beams of your life come from several different sources. They can be from your family values and work ethic, or they can be based upon your political or religious beliefs. It makes no difference exactly where those beams come from as long as they exist. No one can achieve anything of great significance unless it is based upon something that goes beyond the personality of the individual who achieves it.

Then there are the individual rungs. There is a rung at the bottom of the ladder and one at the top. It is important that an individual climbing these rungs not try to jump over one to the other. The analogy I make here is that when you are an entry-level firefighter, the thing you need to concentrate on is becoming the best firefighter you can become instead of worrying about how the fire chief is doing in the fire chief's job. As an individual stands on each successive rung of the ladder, he will begin to gain additional perspective and see further.

Another comparison I like to draw is that good career planning also means knowing how to lock in on the ladder at a given point so that you can get the job done. We all know that by placing our leg over the rung, back through it, and then around the beam of the ladder, we can engage ourselves in such tasks as ventilation, rescue work, and using the tools to gain entry. Locking in on the ladder of life literally means making sure that you stop from time to time to savor and enjoy the function you are currently fulfilling. Many people are surprised to find that their careers have passed them by long before they began to realize the value of the experience. The reason for bringing this point into career planning is that we need to recognize that there are points in time when a person can be satisfied with a particular level on the ladder without criticizing those who are higher up, nor looking down upon those who are lower. The locking-in approach in career planning means respecting the position in which you find yourself and doing everything you possibly can to be the best person at that job.

Our last analogy of ladders is simply the fact that only one person can climb a ladder at a time. There will always be someone ahead of you and someone behind you. The name of the game in career planning is to pay very close attention to the one who is ahead of you on the ladder so that if they make a mistake, you learn to avoid that mistake when you stand on that rung. Likewise if you have ever found yourself in a position at a fire of being on the top rung and suddenly realizing you need assistance, you know that you have to be able to turn around on the ladder and have someone provide you with the tools and techniques to improve your ability to survive. On the career ladder, this means that you need to respect those who are behind you for their ability to keep you where you are.

Well, ladder drill out on the old drill ground probably does not carry with it the gravity of the philosophies I have shared in this chapter. The next time you stand on a twenty-four foot ladder and tie off near the top, take a few seconds and realize that it took a team to get you there but that, at the top of the ladder, you are very much alone. If you have done the job correctly, however, the team is there to continue to support you, and you will be able to walk down from that ladder and go on to another mission and assignment in a real emergency.

In my first fire department, the fire chief really liked wooden ladders. Throwing those wooden ladders was often strenuous, and we never left the drill ground without knowing we had done a job. Hopefully, when your career is over and you have finished with the ladder of life, it can be put back on the apparatus rack with full knowledge that it has done the job.

SELECTED READER ACTIVITIES

1. Write a personal statement regarding ethics and personal behavior for a fire chief.

2. Conduct an inventory of the professional associations in which you have personal membership.

3. Conduct an inventory of your personal professional development library, dividing the list into two groups: technical fire service publications and general management and leadership publications.

4. Develop a list of the projects and assignments in which you have been either the lead person or a contributing member.

5. Develop a list of the specific problems for which you have been an active participant in identifying the problem or in providing the solution within the fire service. Do the same for problems with which you have assisted in your personal life.

MIRROR - MIRROR

CHAPTER 3

Mirror, Mirror, on the Wall:

Meeting Expectations of the Process

This chapter discusses the individual expectations that are created by communities when they are looking for a fire chief. Expectations are reflected in job flyers, job analysis, and performance dimensions. Candidates applying for the job will be compared to these expectations. The emphasis in this chapter is on self-assessment of a candidate's knowledge and skills. This is compared to the dimensions against which the candidate will be evaluated once he has the job. The self-assessment is aimed at helping a fire officer to succeed in being competitive in the promotional process initially, and this is contrasted with the actual knowledge, skills, and abilities that are required to do the job of a chief fire officer after appointment.

Functions of a Chief Officer

As nearly everyone knows, a chief officer has practically nothing to do . . . except to decide what is to be done, tell somebody to do it; to listen to the reasons why it should or shouldn't be done, why it should be done by someone else, or why it should be done in a different way; to follow up to see if the thing has been done; to discover that it has not; to inquire why; to listen to excuses from the person who should have done it; to follow up again to see if the thing has been done, only to discover that it has been done incorrectly; to point out how it should have been done; to conclude that as it has already been partially done, it may as well be left as it is; to wonder if it is not time to get rid of the person who cannot do a thing right; to reflect upon the fact that the person has a spouse and children and family and that they will be affected by the decision, and that any replacement may be just as bad, or maybe worse; to consider how much simpler and better the thing would have been done if the chief had done it in the first place; to reflect sadly that if someone had done it right in the first place it could have been done in twenty minutes, but as things turned out, it took two days to find out why it had taken three weeks for someone to have it wrong. . . . As we said, a chief has nothing to do except . . .

—Anonymous

LEGEND, LEGACY, AND LEADERSHIP

This chapter is going to begin with a test. Tonight, tomorrow, or sometime in the near future, sit down with a friend or a member of your family, someone who is not in the fire service, and ask these questions: "Name three favorite sports heroes." The next question is similar: "Name three famous aviators." Next ask him to name three famous law officers. I bet they can do it easily.

Now for the $64,000 question: "Name three famous firefighters."

Remember, your friend or family member should be an average citizen, not a firefighter. If the person you selected was in the fire profession, like yourself, he might have a few names that would roll right out, such as Alan Brunacini of Phoenix or George Miller of the NFPA. Some older firefighters might remember Keith Klinger, who was fire chief emeritus of Los Angeles County.

Some civilians might remember Steve McQueen as the battalion chief in the movie *Towering Inferno*. There might even be a smart aleck who will recall Smoky Stover, the cartoon fire chief of the 1950s. Two role models that some civilians might name who represented the fire service in the past, both of which were fabrications of a writer in Hollywood, are Johnny Gage and Roy Desoto. These two firefighters, played by actors Randy Mantooth and Kevin Tighe, starred in the television series *Emergency*. A few old timers might be able to recall a television series called *Rescue 8* that was popular in the early 1950s and featured an actor by the name of Jim Davis. Beyond that, no names have emerged to popularize the image of the fire service or the fire chief profession.

The point here is quite simple: Why do we not have any legends in the fire service who are known outside of the profession? This phenomenon does not occur with law enforcement officers, sports figures, cowboys, doctors, attorneys, and a host of other professionals. Personally, I do not believe that Wyatt Earp looked like Hugh O'Brien, and I seriously doubt that Butch Cassidy and the Sundance Kid looked like Paul Newman and Robert Redford. Are all attorneys just like Perry Mason? The icons of Hollywood and television have become legendary in the minds of many people.

In the context of the fire service, images tend to emerge from local perspectives rather than the fabricated ones on television. The image of the fire chief is created and accepted locally. It is a product of the incumbent person's knowledge, skill, and ability. We create images of fire chiefs in our communities, one at a time.

If we continued testing the same group of individuals we tested earlier and said, "Name three big fires," a large number of names would likely come forth, including the great Chicago fire, the Iroquois Theater fire, and the Triangle Shirtwaist fire. During my career, the Interstate Bank building in Los Angeles and the fires of Yellowstone and Oakland might qualify as legendary fires.

Legends, generally speaking, consist of myths, but in almost all cases, they are also based on some kernel of truth. A classic example is the unicorn. From contemporary literature we see that the unicorn was supposed to be mystical and have special attributes. It has been proposed that the unicorn myth probably came from a sailor's description of the narwhal in the Northern Sea. The spiraling horn of the narwhal, combined with a bit of fantasy created by a sailor who was spinning a yarn, could have been the start of the legend. Legends have their place in the imagery of professions, too, but as mentioned earlier, we do not have many of them in the fire service. We do not have a Marshal Dillon, a Dick Tracy, or a Babe Ruth.

The fire service, however, does have legacies. Legacies are different than legends. A legacy is what an individual contributes while serving in a department. It is what the individual leaves long after he is gone. Legacies create the foundation of beliefs and traditions; they are the heritage of organizations. Sometimes they are founded on some fundamental event, but more often they are based on the undertakings of an individual who was accepted as the leader and who became a part of the tradition of the department.

Many of you have probably heard the classic statement about the fire service being two hundred years of tradition unhampered by progress. How can anyone possibly believe that cliché after examining the number of changes that the fire profession has gone through just in the last fifty years? In comparison to many other aspects of local government, the fire service has gone through more major modifications than many of the other so-called public services.

We have gone through many different eras of change. There was the major transition from totally volunteer forces to paid forces in the late 1880s. There was the transition from the man-powered fire apparatus to the technology of the steamer. Then there was the transition from the horse-drawn steamer to automotive apparatus. Probably the most significant change occurred when there was a major transition in code enforcement as fire prevention began to emerge from the catastrophes of the early 1920s. The last few decades have seen the advance of emergency medical services, hazardous materials, and cultural diversity challenges. What is next?

There were problems and new solutions proposed in each of these eras. We talked about that in Chapter 1 when discussing the good old days. In almost all cases, the changes have taken between twenty and thirty years to evolve and be accepted. Most of us have a fire service career that ranges between twenty-five and thirty-five years. This means that unlike many other aspects of our society, most of us only have the ability to take a snapshot of the history of our profession as we proceed along our career path. We are not two hundred years of tradition unchanged by progress—we are two hundred years of change made visible through our traditions.

Doctors, for example, are using some of the most advanced technology available to the human race today. Yet, they all begin their medical career by repeating the Hippocratic oath. This is a fundamental belief that goes back several thousands of years. Repeating this ancient oath does not prevent a doctor from using the most modern and up-to-date technology that is available. Physicians' roots are far, far in the past, but they live in the present. Yet, those in the fire service sometimes accept the idea that the past begins the day they go to work, and the future is the day that they retire. Nothing could be more inaccurate.

Legacies are left as a direct result of the activities of specific individuals who pursue specific achievements during their career paths. For example, there are names that emerge from the history of the fire service that are quite important. Daniel Hayes is credited with being the individual who invented the first aerial apparatus. The concept of consolidated fire protection was the legacy of Chief Keith Klinger of Los Angeles County. Lloyd Layman popularized fog nozzles and gave us a method of attacking structural fires still in use today.

There is one thing that is common in the development of all legends. They do not just happen. Someone worked to create them. Legends do take on a proportion larger than life by accident, but they are created by living individuals who have what it takes to do the job. Legacies are left by individuals who have the potential of becoming legends. Some of you who read this chapter may have an opportunity to fulfill that role in your own department.

Legends, however, do not exist in vacuums. One must perform on the job. This is where the knowledge, skills, and abilities of the individual fire chief come in. This is where the future of the fire service is assured—when individuals decide to take action and to pursue a position that, once it is achieved, will live on long after their career is terminated. This aspect of leadership has three distinctively different dimensions. The first of these is style, the second is vision, and the third is commitment to action.

In short, leadership abhors mediocrity. The only way that an individual makes a contribution is by pursuing something that is different than that which is pursued by everyone else. Style means that a person has a way of communicating, motivating, or building a consensus among a fellowship to achieve a specific objective.

Vision, the dimension of leadership that is most often misunderstood, is based on goals. There are two kinds of visions that leaders devise. The first of these is a positive image, such as changing the world into a better place to live by changing technology. An example of this type of person might be Thomas Alva Edison. The second is the vision to drive people to pursue their individual agendas by changing their behavior—to become better people. An example of this type of person might be Martin Luther King.

Most importantly, vision is a function of principles. The vast majority of individuals who have emerged in leadership roles in organizations have done so based on a desire to accomplish something that is ethically and morally correct. Granted, there are leaders who arrive at a leadership position using the antithesis of these values, but most of them tend to be swept aside in the annals of history.

The last aspect is commitment to action. There is a saying about this dimension of leadership: "Something terrible happens without commitment: nothing." Commitment to action basically means that an individual has the energy, the strength, and the drive to make his dreams come true. For example, very few leaders are captives of the forty-hour workweek. Most individuals who aspire to accomplish something of significance are not clock-watchers. That does not mean they are workaholics; instead, it means that they are focused twenty-four hours a day with regard to what it is they are trying to accomplish.

If we really believe that we are here to make a difference, then we must apply ourselves at all times in the three areas of developing our own style, having a vision, and taking action upon it. Without that type of activity, we will leave no legacy. Most individuals do have a strong desire to leave something as a reminder to society or to their profession that they were there. I once discussed this with a person in the fire service who has now passed on, and I asked him why he felt so strongly about certain issues and why he spent so much time writing on those issues. His answer was relatively straightforward: "I want to make sure that, in the future, every time this particular issue is discussed, my name will be associated with it." Now a scholarship is named after this individual, Chief Warren Isman, who passed away in 1991. The field was hazardous materials.

There is no guarantee that a person who exercises leadership and makes a contribution will become a legend in his own time. Sometimes, because of the emphasis that is placed upon the authority and power that exists around a person in a leadership role, the whole concept of becoming a legend becomes the goal itself. A friend of mine once described an individual fire chief as "a legend in his own mind."

Although none of us are guaranteed that our name will be inscribed on monuments or become a household word, most of us would like to feel that when we leave our profession, we have left something to that profession. It is a form of payback, a form of returning something to an industry that has given so much to us. What your contribution will be is entirely up to you.

This is where the issue of self-assessment comes in. Many people who aspire to become the chief of the department study the topics of fire, firefighting, fire administration, and so on. That study provides a technical knowledge base that is essential to being able to do many of the things that a fire chief must do. Technical competence, however, is a given, not an attribute. A great deal of our training is focused on technology and methods, but little of our training prepares us to contribute to the growth and well-being of the organization. Most of the testing processes that lead to a person climbing the ladder of success follow the technical path. We tend to believe that our technical knowledge is what gives us our status in the community, and we can be seduced by that belief because it is so obvious.

The reality is that few fire chiefs succeed on the basis of technical merit alone. They succeed because they possess some qualities that are just as measurable, but not as obvious. These qualifites can best be described as characteristics, attributes, behaviors, or even habits. As stated earlier, when you are ready to cross the line into a leadership and management role, the ground rules change.

If you read a job description for a fire chief, you will quickly see that technical information plays a secondary role. Most of these job descriptions describe what are called KSAs, which is an acronym for Knowledge, Skills, and Abilities. These KSAs are often fairly broad in nature, for example, knowledge of the management of a fire agency, skills in oral and written presentation, ability to develop good working relationships, and so on. What these KSAs really describe are the things that you will be measured against *after* you get the job.

Read the job descriptions carefully. They are the first clues to whether the things you have done to prepare yourself for the job are the things that management is really looking for. You will likely have to look very hard for specific technical expertise in the job flyers. They may contain statements like "must be knowledgeable in the Incident Command System" or "must be knowledgeable in the Fire and Building Codes," but it is more likely that you will find the flyers full of statements like "must be able to plan [direct, recommend, coordinate, confer, mediate] activities." They are short on specifics, but long on expectations.

Before you decide that you want the top job, you ought to spend a few moments reviewing who you are and what you really believe in. The fire department is not hiring a technician; it is hiring a person to lead and manage the organization. There is a considerable difference. A lot of people who were good technicians derailed after they accepted the position of fire chief. They failed to realize that their departments wanted someone to create a legacy, not to be a legend themselves.

I have read hundreds of job flyers, and in almost all cases, the technical knowledge that a fire officer accrues on his way to the top is secondary to the personal skills and abilities that he acquires at the same time. Knowledge is important, but without the ability to utilize that knowledge, a candidate is vulnerable. The term that is often used to describe the skills and abilities that are desirable is *performance dimensions*.

Performance dimensions are the qualities that often determine a person's ability to actually handle the job. They are not the least bit technical, but they are capable of being defined in technical terms. They can be assessed and measured just like technical knowledge. For example, if the dimension is "writing skills," that dimension can be assessed just as well as the technical skill of calculating hydraulic formulas.

The key here is that if you want to go for the chief's job you should have, as a fundamental base, knowledge of fire science. But more importantly, you should have an accurate and objective view of your knowledge, skills, and abilities in regard to the *behaviors* that you will be required to use in fulfilling the expectations of the job description. Failure to have such a view will create one of two sets of circumstances: either you will never get the job and will never know why you did not get it, or you will get the job for the wrong reason and never know why you fail at it. You do not want to end of in either situation, so let us explore the concept of performance dimensions, and you can answer some basic questions about yourself.

PERFORMANCE DIMENSIONS

With regard to the following performance dimensions, how well do you meet the behavioral definitions? How strong is your possession of the ideal characteristics that separate a mediocre candidate from an excellent one?

Decisiveness

To what extent are you able and willing to make decisions when they are required?

Before you answer this, let me add some caveats: to what extent are you willing to make decisions when there are facts missing, when there is a great deal of pressure or ambiguity, or when the problem is controversial or subject to evaluation by someone higher in the administrative pecking order? A lot of people assume that the fire chief's behavior is focused on fighting fires. It is not. To the contrary, the decisions we make on fires are expected to be in a constant state of revision, and people will accept even bad decisions on fires. But they want a chief to be more decisive in nonfire emergencies. This dimension is measured during discipline cases, code enforcement, budget drills, and inter- and intradepartmental events.

What are the ideal characteristics of a person who is considered decisive? Such a person can make up their mind when presented with alternatives, especially under pressure, without vacillating or procrastinating. In other words, the person is not wishy-washy. Such a person also displays a willingness to make decisions once they have adequate information, rather than asking for more information or procrastinating on bringing issues to closure (known as analysis-paralysis). Decision makers seem to know how to ask the right questions, and they move on a topic when it is right to move. You may find this unusual, but one of the most desirable characteristics of a decision maker is knowing what decisions should be made by others; that is, a decision maker knows how to delegate responsibility. Most importantly, decisive people are not proud of their decision making ability as much as they are proud of the quality of their decisions.

Decision Quality

How *good* are your decisions?

Making a lot of decisions, quickly or not, is not nearly as important as making good ones. An individual who exercises good judgment and considers all the available information in his decision usually makes decisions that will stick. Quality decisions do not have to be redone. Of course, the idea of a "good" decision can fall into the "eye of the beholder" category; what looks like a good decision to one person may not look like a good decision to someone else. But one of the ideal characteristics of quality decisions is that they are supported by others after they are made. Decision quality is often a function of how much controversy is generated by the decision.

This does not mean that all quality decisions are popular. A quality decision maker makes decisions logically and in a manner that is consistent with departmental goals and policies. This may cause conflict because individual employees or groups feel threatened or believe that their interests have been disregarded by a specific decision. A quality decision maker will be able to deal with the reaction to a decision in a dispassionate manner, focusing upon the facts that resulted in the decision rather than the emotion the decision has evoked.

Another ideal characteristic of quality decisions is that they exhibit a depth of action rather than a shallow, impulsive handling of a problem. People who make quality decisions are usually confident enough to defend the decisions without being defensive. They know why they did what they did, and they are willing to say so.

Generally a person who makes firm decisions but allows for alternative approaches if things go wrong or significant new information is presented ensures long-term success for his decisions. Rigidity and unreasonable resistance to changing a decision are signs of poor decision-making ability. Quality decisions also consider long-range consequences and implications rather than focusing only on the short-term immediacy of problems.

Planning

How good are you at planning your activities?

Planning skill is the extent to which an individual can develop a series of logical steps from the current situation to some anticipated future possibility or consequence. Planning is not a piece of paper or a document; it is a thought process of organizing things in a manner that is fairly sequential. Notice that I said *fairly* sequential, because planning is something that goes on continually; it is not like a spigot that you turn on and off. It is something you do to keep moving in a general direction, in spite of the obstacles and interruptions that are placed in your way. General Dwight Eisenhower once said, "In preparing for battle, I have always found that plans are useless, but planning is indispensable."

People with good planning skills anticipate situations or problems and prepare in advance to cope with them. They are aware of both the priorities in the present and the changes that could result from a reordering of those priorities in the future. They are able to anticipate the future implications of alternative actions. The planning process includes the development of realistic goals and objectives and timetables within which to achieve results. Planning also involves understanding and projecting trends and patterns as they impact the organization.

Organizing

How organized are you?

Organizing is often mistaken for planning, but they are different skills. Organizational ability is the extent to which an individual can assemble information, materials, thoughts, and actions into a coherent, orderly, logical unity. Organization has to do with the approach a person takes to a specific problem, and it is evaluated by the way in which a person collects his thoughts and materials in a logical manner. An organized person develops ideas in an orderly fashion, without unnecessary digression or irrelevant information. Furthermore, a person with organizational skills is usually effective in coordinating group efforts.

Problem Analysis Skills

How good are you at analyzing problems?

Problems are as varied as stars in the sky, but problem solving is fairly straightforward. Problem-solving skill is the extent to which an individual identifies and analyzes all relevant facts and their interrelationships while searching for a solution. A good problem solver may or may not be decisive.

When the skills are combined, you have a very effective person. A good problem solver is a person who is objective in seeking and sorting out information. To a good problem solver, pertinent information is more important than volumes of information. Good problem solvers consider the organization's goals and weigh the consequences of alternative solutions and the methods used to solve problems. They also see the relationships between various problems in and among the people involved and within the organization itself.

Productivity

How good are you at completing the tasks you are given?

There is a phrase that is often used to describe the difference between people who talk a good game and those who can play the game. Paraphrased as a question, it goes like this: "Can you walk your talk?" In short, do you do the things that you talk about, or do you just talk about them? The difference is often called productivity. Productivity is the extent to which an individual actively and constructively completes the task at hand, whether in a group setting or as an individual project. Generally speaking, productive people get things done, or they know the reasons why they cannot complete a task. Nonproductive people provide excuses about why they have not gotten things done, but seldom do they blame themselves. People who are productive consistently strive toward meeting their individual and organizational goals. Productivity does not always mean physical work; it also includes contributions of guidance, pertinent information, and ideas to others. Productive people usually are both energetic and enthusiastic, but not always. The most important attribute is the fact that they are effective and efficient. The watchword of the productive person is: get it done! Productive people attempt to complete all work assigned to them within the deadlines that have been set.

Follow-Up and Closure

You can work on some projects forever. How good are you at finishing what you and your subordinates start?

A fire chief is responsible for the productivity of others. That is where follow-up and closure come in. Follow-up is an attribute of a person for making sure that things do not fall through the cracks that have been assigned to others. I once worked for an individual who was so weak in this area that people would play a game with him. The person would make an assignment to his subordinates. The subordinates would put it on the calendar and do nothing. If a month went by and the topic was not mentioned again, the assignment was considered to be forgotten. If the boss mentioned it within the next thirty days, they would mark it down and do the preliminary work but not invest much into the assignment. They would wait another thirty days. If the boss raised it a third time, then something might get done. What usually happened was that very few things ever got to the third mention. More time was wasted in the chief's agenda by this time-consuming activity than by any other problem. Often the effort to not do the work took more time than that which would have been spent doing the job.

There are a couple of key behaviors that successful people employ. First, they make sure that individual subordinates understand the directives and that those directives are carried out, and they set up sufficient controls to ensure timely completion of objectives. Most importantly, they set deadlines that are monitored. A good follow-up person establishes controls up front to ensure that a plan will stay on schedule and reach its objective. Periodic review of program activities and feedback to subordinates in terms of fiscal and program accomplishments are essential parts of follow-up behavior.

A person who demonstrates his own commitment to deadlines is usually more successful in holding others accountable for meeting their deadlines. Follow-up is a necessary behavior, but it can be compromised if the person demanding performance is hypocritical when it comes to his own commitments. If a superior officer expresses a need for control over processes with subordinates but fails to follow through himself, there can be conflict.

Initiative/Self-Direction

Who gives you direction? Do you provide direction for yourself?

One of my bulletin board art pieces states: There are three kinds of people in this world—people who watch things happen, people who make things happen, and people who wonder what happened. Self-starters are the type of people who make things happen. This attribute must be internally driven. *Motivation* is one of the words that is often used to describe the self-starter, but that word has been compromised by overuse. What we are looking at here is simple. Do you need to be told to do something and get punished if you do not do it, or do you choose to do something and take responsibility for it? The extent to which an individual demonstrates self-starting behavior—is willing to originate action without instruction, and is willing to produce new ideas, methods, and interpretations of policy—that person is taking the inititiative. A person who does only what someone else has told him to do, and no more, is in compliance and is good at following orders, but he will not take an organization anywhere.

Self-starters determine outcomes by choosing to act. They do not merely let things happen. Some people describe this as proactive versus reactive. This is more of that management buzzword stuff. Here is the real question: do you seek out opportunity to act, or do you wait until you *have* to act? The answer to that can be a measurement of your inner drive to achieve.

A person with a lot of self-determination will actively try to influence events rather than passively accepting them. Frequently such a person is responsible for developing new approaches, ideas, and methods and for advocating appropriate changes in interpretations of policy. One of the most important attributes of self-starters is that they assume responsibility and control of situations that require prompt action. This is not the same as decisiveness, but it certainly leads to opportunities to be more decisive.

One of the double-edged swords for self-starters is that they may accept additional assignments or devote a lot of extra time to assignments without regard to the impact upon their families and other priorities. The key here is balance and perspective. A self-starting person does not have to be a workaholic. But a person who is not a self-starter will never be in control of his commitments; he will almost always be working on the agenda of someone else.

Significantly, a lot of self-starting people are role models to others. This type of behavior tends to become a group thing. If the person who seeks his own direction works closely with subordinates and encourages similar self-starting behavior among them, the organization tends to be very productive.

Leadership Skills

Are you a leader or a follower? Do you know when to be one or the other?

Everybody thinks he is a leader. In fact, we have overdone the leadership thing in the fire service. We have equated leadership with rank—the more gold you have, the more of a leader you are. Not only is that untrue, it is the source of some of the problems that fire chiefs encounter. *Lead* is a verb, not a noun. You become a leader when you have followers. To the extent to which an individual can effectively direct the behavior of others to accomplish a task or goal without arousing hostility and resistance to change, that person is a leader. Leadership may be demonstrated by virtue of charisma, knowledge, wisdom, assertion of will, or personal style. It is also something that can be shared, and something that should be relinquished to others from time to time.

Leadership is not a permanent position; it is a mantle that is bestowed on those who can carry it. Leaders tend to be people who command positive attention and have the respect of their peers, subordinates, and superiors.

The latter is very important. Leaders often generate the impression of self-confidence, but good ones are never arrogant. They know that they have power to influence others to accept their ideas and to support their positions, so they use this power with discretion.

Good leaders attempt to encourage others to join them, not to *motivate* them. Their objective is to obtain support without alienation. Leaders bring out good performance in their followers by helping them perform at full potential and accomplish their own objectives.

Many words have been spent on this topic in fire service literature. I had one fire officer tell me that as far as he was concerned, he was always the leader of his group because he was given that title along with the title of fire chief. I do not believe that. I think that leadership is an attribute that accumulates throughout one's entire career and becomes a real strength when it is combined with authority to act.

Good leaders are strong without being dominant. They can be equally confident when providing direction and when accepting and responding to counsel, or even disagreement, from peers and subordinates. If a person can encourage the free flow of ideas and yet obtain cooperative resolution of problems with his own goals and objectives in mind, that person is leading.

Interpersonal Skills

How well do you get along with other people?

Everyone has a personality, or at least everyone should have one. A fire chief must deal with a wide variety of personality types. Interpersonal skill is the simple measurement of how effective you are in dealing with people who are different from yourself. Let us put it this way: none of us expects our best friends to act up when we try to deal with them, and yet they do, as do our family members. So what can we expect from total strangers and people with whom we are in competition when we deal with them? Interpersonal skill does not mean giving up your personal point of view, but rather having the ability to deal with conflict with other peoples' points of view without open hostility. The extent to which an individual is able to perceive and respond to the needs, interests, and capabilities of others is the extent to which that individual has interpersonal skill.

Different points of view in looking at the world are the result of different personality types. Your own personality type is not as important as your ability to perceive the perspectives and needs of others with whom you have contact in the course of work and professional interaction. A person with interpersonal skill does not change to please others, nor does he expect others to change to please him. He merely considers differences in selecting and advocating a course of action. The ability to modify your own behavior when you perceive that it is having an adverse impact on others is not weakness; it is awareness.

Establishing, maintaining, and improving working relationships with others is not a simple task. Having the ability to interact, support, and collaborate with work partners is not just a warm fuzzy thing; it is a reflection of a person's awareness that everyone can contribute.

Composure and Self-Control

How good are you at keeping calm under stress?

The one thing about losing your temper is that if you are not careful, you will find it again, lose it again, and continue that cycle forever. Composure is nothing more than the ability to remain calm while those around you are losing it. I call this the "cool school," and I have greatly admired this characteristic in individuals with whom I have worked. The extent to which an individual functions in a controlled, effective manner under stress and pursues tasks to completion despite opposition or difficulties often determines whether that person can remain in control.

Keeping one's composure under pressure is never an easy thing on one as an individual; however, losing it under pressure is a lot harder on everyone else. I have seen a lot of individuals who maintain well under fire ground pressures, yet lose it in administrative situations for the wrong reasons. A person's overall performance should not deteriorate even if there is a mild reaction to stress. Getting angry is not a bad thing, but throwing a tantrum is. One key thing to remember about tempers is that if you must get mad at something, get mad at an inanimate object, not another human being.

There have been many friendships and professional relationships destroyed over a moment of lost control. The attribute of self-control is absolutely essential if you are going to operate in an environment in which there is conflict. No matter how much disagreement there may be, being disagreeable has a price tag that is more severe.

It is normal to be able to accept time delays, constraints, and disappointments without becoming irate or discouraged, it is not normal to blame them on others. If a person's performance remains stable when progress is slow and pressure or opposition is strong, that person is exhibiting self-control. A person who can continue working and not give up under pressure has composure.

Flexibility

How good are you at going with the flow?

There is a country-and-western song that talks about a tree that was strong enough to bend in the wind and, therefore, to survive a storm. The attribute of flexibility is probably one of the hardest to get down pat, for flexibility means compromise in some ways. And being flexible is often looked upon as a weakness. Yet, it is also a strength. The extent to which you can perceive and accept the need for changing your position on a topic because there is some truth to what others are telling you is flexibility. If you modify your behavioral style or position in response to changes in situations or priorities all of the time, you will appear to lack backbone, but modifying your approach to obtain your goals, when appropriate, is effective way of maneuvering around obstacles.

To the degree to which you can shift your behavior to accommodate change in situations without losing your sense of direction, you can be flexible without losing momentum. Rather than utilizing one approach all of the time, which may or may not be universally effective to deal with all problems, you demonstrate flexibility through your resourcefulness in identifying alternatives in working toward your goals. You use a variety of approaches until you find one that works to attain your objectives. This may mean accommodating others when they are uncomfortable in a specific environment. Retreating from a controversial position when you perceive that the facts or the situation has changed or that your group support has changed or evaporated is a sign of flexibility.

Flexibility does not mean giving up or turning completely around. It can look like reversal, which is what makes this attribute so difficult to assess at times. Just remember that a flexible tree or rod that is bent in a certain direction will, as soon as the pressure is released, return to form. That is the second aspect of flexibility—going back to what you are trying to achieve, not abandoning it, is the goal of being flexible.

Oral Communications Skills

How well do you communicate in private conversation and in public presentations?

Everyone knows how to talk, but not everyone knows how to communicate. The extent to which you can express your ideas clearly, concisely, and persuasively in individual or group situations is the extent to which you have good communications skills. Communication includes listening skills, body language, vocabulary, grammar, and a host of other skills. Being able to utilize nonverbal communication is almost as important as being able to speak.

An individual who can listen to others and bring them into a conversation is a better communicator than one without that ability. Listening attentively to what others have to say and asking them relevant questions is almost as important as communicating your own ideas.

Speaking clearly and concisely is not to be taken for granted. If you think that you do this effectively, I suggest that you either listen to one of your conversations that has been tape-recorded or watch yourself on a video-tape. You may be amazed at the number of incomplete sentences and of "uhs" that occur when you are speaking.

One of the quickest ways to assess a person's speaking skills is to directly ask the person a question or ask them to explain an issue and then sit back and listen. A good communicator will answer the question without confusion, disorganization, or rambling. A poor communicator may use a lot of words but say nothing.

The fire service is often criticized for "talking shop" when communicating outside the fire service. We have enough acronyms to fill a book, and so do the other professions. Sometimes we do not know what they are talking about, and they do not know what we are talking about. Use "fire-manese" within the firehouse, but not in the nonfire world. A good communicator will translate technical language to the audience's level.

There are essentially two different levels on which you should assess your communication skills. The first is the personal or informal level, and the second is the organizational or formal level. You need to be able to function at both levels. Your presentation of ideas and your poise and persuasiveness all figure into the image you will create as a fire chief.

I often hear people say, "I am not a public speaker." That is not a good excuse for not being prepared. It is a cop-out. Public speaking is not a god-given gift, and a lot of people who do not like it are very successful at it. What I am suggesting here is that you understand that oral communications skills are an important part of how you will be assessed by others. Your communication skills are often used as a measurement of how much control you have over yourself and the various situations you encounter.

Oral skill also includes an assessment of how sensitive and responsive you are to the differing communication requirements of those to whom you speak. Speaking down to people or ignoring their needs is not a good way to influence them.

One of the more interesting aspects of communications skills is the need to assess when to shut up. It is all right to discuss topics thoroughly, but if you lose sight of the need for clarity and go on and on, you can go over the edge as a communicator. Avoiding lengthy explanations and comments can be an asset in communication.

Written Communications Skills

How good are you at communicating your thoughts in writing?

This is an area where the fire service, in general, is very weak. The skill of writing is one that requires practice. There are many people who can communicate fluently when talking about a subject but who seem to go brain dead when they pick up a pen or face a keyboard. This is a serious deficiency in the fire service, for a great deal of the assessment of fire service work is done through reviewing reports, correspondence, and documents. By assessment, I do not mean the way in which we evaluate ourselves; I am talking about how we are evaluated by city managers, city attorneys, elected officials, and even court judges.

An individual who can express ideas clearly in writing can be more successful in the long run. Good grammar is as important as a high level of technical knowledge. A good vocabulary is as important as having a lot of facts. Clarity of written documents should not be impaired by problems of spelling and punctuation.

When a document is completed, it should contain all of the significant facts, but the amount of detail should not be so elaborate that the document is confusing or rambles on unnecessarily. One skill that is needed in written communications is the ability to edit written documents. You should be able to recognize the strengths and weaknesses of your documents upon review, and you should also be able to recognize the strengths and weaknesses in the written communications of your subordinates.

Awareness of Community/Social Issues

How much knowledge do you possess of issues outside the fire service?

A fire agency exists within a context. Usually that context is a jurisdiction, whether it is a city, a fire district, or a county. Within that jurisdiction, there are people, homes, businesses, and issues. It is an environment that must be accurately understood if a fire chief is to serve the community well. This attribute involves an assessment of how well-connected a person is with what is going on within the entire community. An individual needs to be able to demonstrate awareness and understanding of the interaction between the agency's organizational goals and duties and the needs, interests, and social structure of the community.

At one level, this awareness involves an appreciation of the importance of public relations in departmental programs and services. At the other end of the spectrum, this awareness involves an appreciation of how departmental decisions will be either accepted or rejected by the community's special interest groups, or the "power elite." Social, racial, and ethnic prejudices that are incompatible with the community are part of social awareness, too. A socially aware person can be fair and objective with all people.

In the past, there have been chief fire officers who have achieved a great deal during their time in office. Almost all of them did so because they possessed a quality or a combination of qualities that allowed them to succeed where others might have failed. A person who is anticipating the possibility of following in the footsteps of these chiefs must evaluate their own qualities and either accept them as they are or actively seek to improve them.

SELECTED READER ACTIVITIES

1. Visit a local bookstore and obtain a copy of a recent book on the subject of personal self-assessment. Over time these titles change, but a book like *Please Understand Me,*[1] by David W. Keirsey and Marilyn Bates, or any current text about the Myers-Briggs type indicator, is useful. Read about the concept of personality development, and perform your own self-assessment.

2. Review the list of attributes mentioned in this chapter, and evaluate your strengths and weaknesses with respect to them.

3. Discuss your evaluation of your attributes with a close friend or mentor to see whether they agree with you on your self-assessment.

4. Develop a list of scenarios or situations in which you have demonstrated these attributes. Critique your performance, and develop a list of things that you can improve upon without outside assistance.

5. Seek every opportunity to make public presentations. Tape- or video-record them, and then listen to them or watch them. Eliminate any bad habits.

6. Seek every opportunity to prepare written communications. Edit your own work ruthlessly, and seek opportunity to edit the work of others. Practice writing skills at every possible opportunity.

---

[1] David W. Keirsey and Marilyn Bates, *Please Understand Me: Character and Temperament Types* (Del Mar, CA: Prometheus Nemesis Book Co., 1984).

WHAT THEY DIDN'T TEACH YOU

CHAPTER 4

What They Never Teach You at Fire Academy:

What You Have to Gain and Lose by Becoming Chief

This chapter focuses on the price one actually pays for being selected for a leadership and managerial role. The chapter includes an exploration of the relationships between the candidate and the people who hire them, and candidates and fellow department heads, family, peers, and friends.

> *I have nothing to offer but blood, toil, tears and sweat.*
> *—Sir Winston Churchill, First Speech as Prime Minister, 1940*

IS THE CITY MANAGER THE GUIDING LIGHT?

A city manager, who was also a friend of mine, minced no words when I asked her what she thought about the relationship between fire chiefs and city managers. She said, "It is highly unlikely that many city managers care to or try to mentor their fire chiefs to become a more effective member of the management team." When I asked her why she thought this was the case, she answered, "Because most city managers look upon the fire chief as being an alien to the overall management process."

This conversation occurred in a meeting with the city manager concerning the need for improving the quality of training for fire chiefs. This particular city manager was a strong supporter of the fire service and, over the years, had been actively involved in developing personnel to serve the fire service. Yet, the cynicism reflected in that conversation was real.

The fire service needs to accept the fact that the average fire chief is not looked upon by the average city manager in a favorable light—a generality that may not fit every specific situation. This generality weighs heavily on the fire service in ways that people have yet to identify. Moreover, the entire concept of fire protection may be restrained because of this relationship. So, what are we going to do about it?

It does not do much good to complain about the fact that people in city management have lost their sense of respect for the fire service. Fighting back by stating that we do not have much respect for city managers, either, does not help, since that puts us into the mode of the win-lose proposition. Unfortunately, the hierarchical relationship between fire services, city management, and political structures is such that the fire service almost always comes out on the losing end.

We could choose the easy way out. What if we just admitted that we are not doing all that well in the management structure of government and relegated our profession to a second-class, blue-collar, nonprofessional status? We could abandon the field and let some other profession take responsibility for fire service leadership.

This is something they do not tell you at the Fire Academy: we are not considered to be top-quality managers, nor are we looked upon as being on the leading edge when it comes to anything except fighting fire. Perhaps this explains the popularity of the concept of public safety. In that same conversation with the city manager, she said that, generally speaking, city managers regard the choice of police chief as one of their highest priorities in order to maintain stability in their governmental situation. This brings us to the point of the discussion in this chapter: what do we need to do to bring the fire service leadership into a position of power and persuasion that is parallel with that of our brother agency, law enforcement?

It is not going to be easy to attain a separate but equal status, but I am convinced that it can be done. As a matter of fact, I would say that even though the cards are stacked against the fire service, we have some skills and abilities that others lack. If we recognize the fact that a strategy is essential for the long-term survival of the fire service, then we must embark upon a plan to bring about change.

The first thing we need to do is to step back and look at the preparatory stage, at what makes the difference between a police chief and a fire chief. How many police chiefs do you know who attempt to take over every incident that occurs in the street? How many police chiefs do you know who are really afraid to deal with controversial issues? How many police chiefs do you know who are not involved in community affairs?

We need to spend some time looking at our peer professions and determine the similarities and the differences of those professions and our own. Frankly, the qualities of police chiefs implied by the questions in the previous paragraph do not make them any better than fire chiefs, but if we stand back and take a look at what makes police chiefs successful, we will find that they often have skills, competencies, and abilities that are different from those that we exercise in the fire service.

A lot of people will object to this statement, but police chiefs, for the most part, are much more highly educated than fire chiefs. There are good reasons why this is the case. In the first place, a lot of money has been dumped into law enforcement over the last couple of decades to pay for that training and education, especially by the federal government. But that is not the only reason. In general, by the time people have worked their way up through law enforcement, they have almost always availed themselves of mainstream educational opportunities; they are better read and have a more universal understanding of society and societal problems than the average fire chief.

The second step we need to take is to start developing skills and abilities that have nothing to do with fire protection or fire technology, that is, competencies that are more universal. I readily admit that the decision making that goes on at the scene of the fire is highly stressed and highly technical, and that there are many people in our society who could not do it.

Yet, this experience has created a "ready, fire, aim" approach by which the fire service is often derailed when dealing with complicated problems. The last chapter described areas of behavior that will enable almost anyone who is good in those areas to do well at anything they undertake. Your self-assessment in that chapter may have shown you areas in which you need remedial work; if not, ask someone else to score you and tell you what they see. It is time for us to be more realistic about our strengths and weaknesses.

In general, fire officials tend to be less articulate in public settings and have a limited ability to put their thoughts into writing in comparison with other department heads. There are a few individuals who are articulate and eloquent, but they are the exception rather than the rule. The development of the kinds of competencies that are needed requires the fire service to stretch itself far beyond the culture of the station house and headquarters. It requires reaching out into the community and putting one's own reputation at risk by trying things that have nothing to do with the fire service. It means getting involved in projects that are much more complicated than day-to-day activities.

The third step we need to take is to start building bridges with city administrators. I am not talking about coffee klatches and good old boy/girl networks, I am talking about fire chiefs developing a stronger bond between city managers and themselves by voluntarily joining the team to assist the managers in getting *their* job done.

This step is filled with terror for many people. One of the common myths is that by stepping forward and joining the management team, you are abandoning your own roots. When I talk about joining management teams, I am not suggesting that we suddenly disavow our relationship with the fire service but, instead, that we offer to use the fire service as a resource to solve community problems.

You might as well get this news now rather than later: when you take on the job of fire chief, you are going to lose a lot of old friends and be forced to make a lot of new ones. You can hang onto some of the old friends if you have prepared yourself correctly, but you will be in real trouble if you do not make some new friends in the job of being chief.

There is a tendency for fire chiefs to alienate themselves from city management by making sure that their working day and their working relationships are totally away from city hall. They do not like to participate in activities that involve other department heads. Generally, fire chiefs prefer to stay amongst their own kind. Unfortunately, the comfort zone with which we surround ourselves in the firehouse often insulates us so much that it prevents us from having a good working relationship with the city manager and the other department heads.

The fourth change we need to make is to challenge the status quo. If we go back to our police chief compadres for a few moments, you will probably note that many top law enforcement officials have developed their names and reputations around developing concepts, for example, community-oriented policing, the SWAT team, neighborhood watch, DARE programs, and so on. It is unlikely that any city manager is going to have a great deal of respect for the managerial skills of an individual who is basically just doing housekeeping.

There are limitations to this, of course. One cannot go around changing everything all of the time and hope to enjoy a reputation of being an agent of changes. Common sense, focus, and a strong sense of pragmatism go a long way toward taking a concept from theory to reality.

The fifth step is: compete. Broaden your perspectives. Compete for fire chiefs' jobs in larger communities. Make it your priority to learn some kind of lesson every time you enter a competition. Compete for assistant city managers' jobs. Compete for professional leadership roles. Compete every time there is an opportunity to demonstrate the difference between a mediocre individual and a highly competent one.

It is probably true that city managers are not going to pay a lot of attention to this particular component of the process. In some cases, they really do not care. In other cases, competition may cause you some grief because it will expose you to assessment by others, and there is always the possibility of failure. Compete anyway!

Competition brings out the best in all of us. Individuals with average skills begin to show improvement when they are up against individuals with even slightly above average skills. In competitive situations with individuals who are highly skilled, the more highly skilled person never loses anything, but the person with average skills may show drastic improvement after only a short period of time.

The sixth step is to get a mentor. If the city manager who talked to me was correct, most city managers are not going to look amongst fire chiefs for someone in whom they can invest some time. However, this does not mean that there are no city managers willing to help fire chiefs who ask for assistance. So, recruit upward. Ask for guidance and counseling, and even seek additional job opportunities, by going to city management and asking for assistance.

Do you know what will probably happen when you go to city management for assistance? In some cases, you will be turned down. In many cases, you will be given excuses about why something cannot be done based upon the level of cynicism expressed in the opening comments of this chapter. But sooner or later, you will be given an opportunity. It may be a small one, such as chairing a committee, being in charge of a task force, or taking on an undesirable job. The consequences of your success may not be chalked up as a major achievement for the community, but two things will occur. First, you will have obtained some insight into yourself as a manager and leader in the generic sense, and second, you will have either reinforced or disputed the traditional perspective that fire chiefs are too narrowly focused.

An interesting question was raised in a workshop one time. A city manager was attacking the fire service in a somewhat abusive manner, going on and on about how he felt that fire chiefs did not measure up in this way and that. One individual in the audience finally had the courage to stand up and ask, "Who is it that hires these fire chiefs anyway?" The city manager had to admit that most fire chiefs are hired by city managers. The individual in the audience then questioned who is really responsible for the current state of affairs. Is it the fire chief for not being adequately prepared, or is it the city manager for accepting individuals who do not live up to the expectations?

The jury is still out on the issue, but some observations can be made that lead us back to the original contention of this chapter. First, the city manager who selects a person who performs below average does a disservice to the community. And second, those individuals who are competing for fire chiefs' jobs but who are unwilling to set their sights higher in regard to their own range of skills and competencies are doing the profession a disservice. If these two groups are allowed to continue perpetuating the myth that fire service personnel make poor managers and leaders, the consequence is a diminished profession.

The problem will not be reconciled until city managers state their expectations that a fire chief be no more and no less a member of the team that operates the community than the police chief, the public works director, and the finance director, and until we in the fire service accept the challenge of raising our own sights and recognizing that some people will fail. Raising our own sights will go a long way toward raising the credibility of the profession.

BE CAREFUL WHAT YOU WISH FOR

It appears that those who seek fire chiefs through the testing process look for perfection. All you have to do is read the job descriptions to discover that they are looking for people who are progressive, innovative, creative, and charismatic. As you read some of the job flyers, you almost feel compelled to add, "can leap tall buildings in a single bound and stop a speeding locomotive."

Of course, when we read those job flyers, we begin to subscribe those characteristics to ourselves. After all, are we not progressive, creative, innovative, and charismatic? Generally speaking, candidates skim over the rhetoric of these items and look for more substantial indications of the job, that is, pay range, size of the department, and perks.

Then there is the testing phase in which candidates are pitted against one another to determine which person is best for the job. A group of assessors or evaluators come in, furrow their brows, place pencil or pen on paper, and try to describe which person is best for the particular job. Nothing is more satisfying than coming out on the top of one of these lists. One cannot help but feel that if one is at the top of the list, one must possess all the necessary characteristics.

Once you get the job, there is a honeymoon period. During that time, everyone gives you the benefit of the doubt, and no matter what your personal characteristics are, you can usually get away with making some changes that may or may not have long-range implications. During the honeymoon period, everyone is getting to know one another. All honeymoons, however, must come to an end.

That leads us to the next phase—the shock of reality. Many individuals have expressed dissatisfaction with the fact that they tested for a position that ceased to exist once the honeymoon period was over. Often, in spite of the rhetoric of the job flyer, the rug is pulled right out from beneath innovative, creative, progressive, problem-oriented, charismatic fire chiefs. In short, many of them are told not to be quite so progressive, quite so innovative, or quite so creative. The very dimensions that allowed them to seek the job and to achieve the position are often a source of great dissatisfaction once they are expected to address specific problems in the fire service.

The consequences of this contradiction are multiple. On the one level, the fire chief becomes increasingly dissatisfied that all those great things he wanted to do are put on the back burner. Secondarily, there is often conflict between the chief and the appointing authority that results in strained relationships. Lastly, there is the loss of energy in the organization because the status quo seldom requires much energy to sustain.

A group of chief officers once told me that they felt a strong sense of betrayal because of this phenomenon. During the recruitment and selection process, they were encouraged to express their ideas on how to improve the fire service. Yet, after appointment, they were told in many cases to temper their enthusiasm, slow down, do not make waves, and so on. Several individuals indicated that this probably caused more stress in their jobs than almost any other single factor.

So, what is a candidate to do? Are there any techniques to make sure that a chief officer is not romanced into a position that subsequently results in a divorce? There appear to be some common factors that are shared by those people who have had positive experiences of going in and making changes in their organizations. These factors are (1) a sort of prenuptial agreement between the chief and the appointing authority, and (2) a recognition that openness and honesty in the chief's relationship with the appointing authority is by far the best policy.

The use of a "prenuptial" agreement allows a contract to be established between the fire chief and the appointing authority. In the writing of a contract, a candidate has the opportunity to articulate expectations. These expectations can encompass anything from salary and benefits to a statement regarding the candidate's fire protection philosophy and agreed-upon performance standards.

Top-level athletes have been using contracts for years to establish the ground rules for performance. There is nothing wrong with fire chiefs having the same working relationship. A contract is a legal document that provides protection to both parties. It can be used by the appointing authority to indicate to an individual that he is not meeting expectations. On the other hand, it can be used by an incumbent to delineate what is to be achieved within the contract period; if the appointing authority reneges on that agreement, the chief is justified in seeking redress.

One step down from the contract is the concept of a Memorandum of Understanding. This is nothing more than a statement exchanged between two parties to clarify specific points prior to entering into a working relationship. Some individuals have been successful in getting their appointing authorities to put their expectations in writing. This goes further than what was contained within the job flyer. It specifically addresses what the appointing authority wishes the chief to accomplish within the context of the organization.

A Memorandum of Understanding is not nearly as binding as a contract. In one sense, it is nothing more than a statement of intent. However, it has one distinct advantage over the contract. It can be modified as time goes on by dialogue between the chief and the appointing authority. It is not nearly as restrictive as a constract, and the consequences of failure on either part are not as visible or controversial.

With either of these techniques, the agreement must be established prior to appointment. One cannot negotiate an agreement after one has accepted a position. That is a position of weakness. Many candidates are fearful of entering into an agreement primarily because they are afraid they will offend the appointing authority, and that may be true. On the other hand, the degree to which an appointing authority reacts to this issue may well portend how things will go after the candidate has been appointed. Beware of an appointing authority who is unwilling to talk about actual job expectations.

A lot depends on the chemistry between you and the person who is asking you to accept the appointment. There is nothing inherently wrong with asking for a clarification of the major issues. A simple question you can ask during the selection process is, "Does this agency have any difficulty in entering into contract for this position?" City managers, for example, often have contracts for their services, so there is no reason fire chiefs cannot have them.

There is another reason for trying to get an agreement clearly articulated before you accept a position. It is not uncommon for appointing authorities to bring someone on board, and then depart. More and more, with the stresses and strains of government today, appointing authorities bring people into the team environment without adequately exposing them to conflicts that exist between themselves and the political bodies. You might be surprised when you ask for clarification of job expectations to find that the person responsible for making the appointment will start to divulge information about political stability, or the lack of it.

I personally observed a worst-case scenario in this matter. A department head had resigned from his old job and was waiting to come to his first day at the office in his new capacity. He was asked at this point to attend a council hearing and listen to some budget deliberations. During the budget deliberations, one of the council members made a motion to delete the position that this candidate had just accepted. The individual almost had heart failure. Later the city manager indicated that he knew the council member was going to make the motion but that he knew there were three votes in the candidate's favor. This situation might have been either humorous or avoided altogether by some dialogue about expectations and the environment in the organization prior to accepting the position.

Nothing in this chapter will prevent an individual from seeking a job that ultimately results in frustration. But remember, if the authorities are asking for a person with superdimensions and you possess those dimensions, make sure that when you actually accept the job, you will be able to apply yourself. Otherwise, you will only be setting yourself up for a great deal of personal grief.

POSITION POSITRACTION

If there was one event on television that I never understood, it was the tractor pull. The idea of a whole bunch of trucks trying to haul a big, heavy sled over very short distances did not appeal to me, that is, until I actually saw one for myself. With these huge trucks belching diesel and roaring like sheet metal dinosaurs, I could not help but be impressed by the sheer energy of it all.

I noticed an interesting phenomenon at the tractor pull. Even with all that power and energy, if the vehicle lost traction and spun its wheels, it would almost always end up being a loser. Those that kept traction and were pulling with each revolution of the wheel ended up being winners.

Individuals, organizations, and you, too, can lose traction. While expending a great deal of energy and making an awful lot of noise (like tires spinning on the pavement, which is the epitome of attracting attention while doing absolutely nothing), they go nowhere.

How do you know whether you will have traction? Is it possible that you are going to be spinning your tires? Well, there are a few tests you can use to determine whether, given your opportunities, you have maintained positive traction.

The first step in this process is to take a look at the past. You need to compare where you are today with where you were last year, maybe even during the previous three years. If things are pretty much the same and very little has changed, you have no traction. This is not unlike the situation in the Bill Murray movie *Groundhog Day,* where he had to live the same scenario every day, over and over again. Nothing ever changed, until he changed. Then life went on.

Another litmus test is the reality check. The reality check consists of asking other people about how they perceive your growth and development. I am not talking about asking your friends whether they still like you; I am suggesting that you talk to the people who are evaluating your performance.

Taxpayer groups are becoming increasingly hostile toward fire and law enforcement. We need to look at them as customers instead of competitors. After all, they are the ones who pick up the tab on the services that we provide. Talk to some of them about your performance.

It is incredibly difficult for the fire service to analyze public sentiment. There is a tendency for those of use in the fire service to believe that we still wear the white hats. As a matter of fact, it is common practice, in most fire organizations, to shy away from controversy for fear that it will offend the community. Interestingly enough, law enforcement seems to embrace controversy to a large degree, and they are not taken for granted nearly as often as fire protection services.

Another way of assessing your personal traction is an internal audit of your current position. While the fire service has, over the last couple of years, embraced participative management more and more, there are still serious deficiencies, and in some cases, there has actually been a backlash against participative management in the light of current financial crises. It is very difficult for people to sit down and talk with labor groups when it is anticipated that there may be conflict over forthcoming decision-making processes.

Yet, that is all the more reason why you need to do internal audits. If you ever have driven a pickup with no weight in the back end under adverse circumstances, you realize that no matter how much you rev the throttle, unless there is something in the back that is weighing the vehicle down, you will not get any traction. So it is with organizations. No matter how much we rev the throttle up front in the fire chief business, unless we have someone who is bearing down in the back, we will not go anywhere.

This component of assessing reality is very difficult for some people. The real reason it is avoided is that it often reveals communication problems and lack of leadership at the top. It takes a courageous organization to participate in this kind of process. There will be those who will attempt to sabotage it on both ends. Some management personnel do not want to accept the fact that their subordinate personnel have anything to say about the way an organization operates. Some labor personnel do not want to agree with any direction the organization is going unless it is to preserve the status quo. Both attitudes are equally fatal to the ability of the organization to move forward.

The last test for traction is also very difficult. It is called the internal check. Some call it intuition; I call it the "gut-level feeling." No matter what terminology is used, it boils down to you, as a fire officer, asking yourself the tough question, "Am I and this organization going anywhere?" If the answer is a resounding yes, you should be able to produce a considerable amount of evidence from the previously suggested litmus tests. If the answer is no, then the responsibility now falls upon your shoulders to start doing something about it.

How do you get traction back if you have lost it? Sometimes it means slowing down just a little bit, placing yourself in a lower gear. It means not applying as much energy to superficial activities but making sure that you are moving forward, even if it is more slowly than you anticipated. It may even involve stopping long enough to get outside the system and look back at it to see what is causing you or your organization to lose its traction.

If your vehicle has ever been bogged down in mud or snow, you know that there are times when you have to get someone to come along with a tow truck and pull you out. There is nothing inherently wrong with seeking assistance from the outside to get traction back. There are several places from which this can be obtained. The most readily available source is your peer group of fire officers. Professional associations over the last decade have become more interested in helping individual fire officers succeed. They have formed informal networks that can come to the aid of an individual and help him by providing some insight that only an outsider can provide.

A second hypothetical tow truck could be the training and education system. If you are spinning your wheels, perhaps it is time to focus on your most recent training and educational opportunities. It is unfortunate, however, that departments generally cut their training budgets at the outset of a financial crisis. This is probably one of the most unwise strategies anyone could employ.

The last tow truck could very well be your own boss. There is usually a great deal of personal support there. It takes a lot of courage to admit that you are floundering a bit, yet the very admission of that reality may be the key to your future success.

The corporate world of business uses three phrases to describe this need for renewal: takeover, make over, hand over. It is not uncommon for a CEO to come into an organization that is facing difficulty and potential oblivion and to relish the fact that things are difficult. You do not hear a great deal about individuals who are managing organizations that experience little difficulty, but there are entire groups of people who have built their careers upon the fact that they know how to take over an organization and then make it over so that it ends up being more viable. This mentality involves determination and a high degree of self-esteem, but, when it is employed in conjunction with your superior, it can become an extremely powerful set of circumstances.

Individuals with a lot of traction are like the little engine that could. They can pull heavy weights over extremely high obstacles, and they always succeed. Individuals who lack traction are full of excuses for their own failure. Which of these two would you prefer to be?

MR. OR MRS. CHIEF

You just decided to go for the promotion. You are so proud of yourself. You expect that you will come out number one on the list and that the city manager or fire district board will offer you the fire chief's job. You are going to have to move to a new town, but that is no big deal. Your moving expenses will be paid, you have a contract, and you know what to expect. But when you go home and make the announcement to your family at dinner, instead of unbridled enthusiasm, you are greeted by silence. What is happening?

When you get promoted to the rank of fire chief, your entire family goes along whether they want to or not. In the process of developing people for future promotion, we seldom spend much time talking about the impact and consequences on spouses and children when they ultimately achieve the job. The problems associated with relocating families and some of the consequences of moving from one department to another are not trivial. I suggest that you focus attention upon the impact on the immediate members of your family when you aspire to take the position of fire chief.

At one level, the prospects of promotion result in euphoria. After all, a promotion usually means an increase in salary, perks, status in the community, and a whole host of things that have favorable inferences. On the other hand, a promotion to fire chief often means loss of overtime and flexibility in your personal and family schedules, and an increase in visibility for practically everything that you do. It may even change your circle of friends and acquaintances. The process of promotion carries with it the implication that the family now has to shoulder some of the responsibilities that you will soon raise your right hand and swear to uphold.

The wife of one of my former chief officers used to refer to my wife as Mrs. Chief. Then my former subordinate became my peer when he became a chief in another community, and true to form, my wife reciprocated by calling his wife Mrs. Chief. I listened to this dialogue for several years before I realized what they were saying. We get the badge; they get the problems of making a family work with the job.

Anybody who is married to someone with a certain degree of visibility or a leadership role is often placed into a form of anonymity behind that title. There are notable exceptions to this, of course, where spouses are of an equal stature either professionally or personally. In any case, the individual who is married to a person who has aspired for a leadership role may or may not have any interest in the function of that leadership role.

I would defy anyone to show me where we in the fire service have an effective training program in place to prepare a spouse for being married to a fire chief. It is strictly on-the-job training. Our spouses prepare for that capacity merely by listening to us, when we come home, either explain what is going on in the organization or complain about it a lot. The vast majority of a spouse's preparation for becoming the spouse of a chief consists of being the sounding board for all of the problems that the person goes through in attempting to achieve that goal.

Our spouses are not sent off to the National Fire Academy for a couple of weeks at a time to learn methods, techniques, and shortcuts to use in dealing with us. Instead, they remain behind while we are off at these training and educational events. In many cases, they are required to take on a lot more responsibility than a normal spouse because of our absences as firefighters. The wife of a fellow fire officer once told me that she had become a minor mechanic for fixing things around the house because, as she stated it, "My husband is never here."

Further, these spouses often put up with a lifestyle that we in the fire service accept as normal, but it is a style that would be considered extraordinary by most people. Do you know many people whose spouses are often torn out of the bed at three o'clock in the morning in response to a pager call or a ringing telephone that results in them hurriedly dressing, rushing out the front door, and charging off in the general direction of something that is potentially lethal? Even law enforcement officers have a certain degree of insulation between their private lives and their professional duties. I know of very few law enforcement agencies that recall their personnel under stress conditions as frequently as we do.

We who wear the badge of a fire agency take this lifestyle for granted; we accept it with little or no concern because it comes with the job. On the other hand, our spouses, who do not have badges, have to cope with the consequences of these disruptions to their lives.

I do not want to make it sound like a fire chief's lifestyle is all bad; it is not. But when a person becomes chief, the family often goes through culture shock. When we are working shift work, we have a lot of spare time to devote to family projects, special events, travel, and so on. Once a person becomes the chief, a lot of that goes away. And then there is the phenomenon of the chief suddenly being home every night or, worse yet, being gone every night to meetings. The chief may not be available to do things during the daytime that previously had been routine.

The fire service has a very high divorce rate among individuals who move up through the rank structure. Although I do not have nationwide statistics, in my own anecdotal way, I have collected information regarding the divorce rate among fire chiefs. It is actually a rarity to find a fire chief who is still married to his first spouse. I cannot prove that the stress of the fire chief's job itself is the basis for this phenomenon, but conversations with chiefs and their spouses have revealed that quite frequently first spouses were burned out in the process of competition that it took to prepare for the top job. Second and subsequent marriages are usually to individuals who have entered the picture without a lot of the baggage of the competition process from day one. In short, a second spouse may have a different perspective on being married to a chief than a spouse who sees the chief start from firefighter and work through the ranks.

In thinking about spouses and families of those who aspire to become chief, I started contemplating strategies that one could use while developing oneself for the top job. There are probably no ground rules that will apply to all people at all times, but there are a few things that we can be thinking about that should help.

First it is important that individuals who are aspiring to a fire chief's job openly engage their spouses in a discussion of the changes that the job may bring to their lifestyle. I am not talking about increases in income. I am talking about decreases in flexibility and increases in the pressures of the job. These are not things that you, the badge carrier, are going to be able to shoulder entirely on your own. This dialogue may not be easy because of the fact that you may not know what the expectations are until you get into the job. In my experience, there is a spectrum that ranges all the way from an expectation that spouses are to be left out of almost all aspects of the chief's job, to the point where the fire chief's spouse is expected to be the chief's social and professional companion in a wide variety of political and interpersonal settings.

It is important that we take into consideration our spouse's personal feelings and concerns about these expectations. If we go into a job in which our spouses are to be excluded with companions who wish to be included, we may experience some social pressure. Conversely, if our spouses are supposed to be part of the fabric of the job and our companions are basically shy and retiring and do not wish to be involved, there can be some pressures on us.

Part of this strategy should involve a full discussion with the appointing authority when a person takes a job regarding the expectations for both the chief and the chief's spouse. The idea that there is any correlation between our spouse's performance and our own performance is considered repugnant by most. We do not wish to be held accountable for this kind of relationship anymore than we want to be held accountable for our ethnic backgrounds or educational liaisons or personal tastes. Yet, the reality is that by not taking a look at these relationships, we may be creating stresses in them later.

A secondary strategy is to share information with your spouse about the positive side of the job as opposed to sharing only complaints. I am amazed at the number of chief officers who do not take home and share with their husbands or wives any of the documentation that crosses their desks about the positive aspects of the fire service culture. Granted, many individuals provide their spouses with this experience and exposure by taking them to conferences and workshops. However, there is a cornucopia of information that crosses our desks almost daily that can assist our spouses in understanding the culture of the fire service. I am referring to things such as the IAFC's newsletter "On-Scene," the professional bulletins that we receive, and even the fire service magazines. I have been surprised by the number of spouses who have told me that they read *Fire Chief* magazine. As a matter of fact, I probably have received more feedback from spouses regarding certain columns that I have written for that magazine than I have received from their fire chief spouses.

A third strategy that could be considered in this area is to make sure that your spouse is involved in the actual decision to accept a job. Although the control may be difficult for some people to relinquish, the decision to accept the job when you are married is not just your decision. This strategy often involves talking through issues that may result in feelings that the job is unacceptable. I knew a fire chief who was given the opportunity to move to a new city. In the course of this transition, he was going to get a significant pay raise. After engaging his family in a discussion, however, he declined the job. The reason had nothing to do with his professional competency, but with the fact that two of his children were in their last two years of high school and had been in the same school system for almost all of their teenage years.

In this case, the chief officer was an individual of a very high level of competency, and I am sure that he will be given another chance sometime in the future to accept another job. What I saw in his decision was a great deal of maturity and wisdom in placing his family's values equally as high as his professional ones.

Spouses are important. As chiefs, we need to consider that our job security is not only a case of our performance rating by our superiors but also of our performance rating by our loved ones.

SELECTED READER ACTIVITIES

1. Conduct an interview with a highly regarded city manager, or a district or county administrator, to determine his perceptions of what a fire chief role should be.

2. Review the job descriptions you collected in Chapter 1 and the self-assessment you did in Chapter 3, and compare your capabilities to the expectations stated in the flyers.

3. Develop a mentor to discuss your positioning and preparation as a candidate.

4. Prepare a list of the pros and cons of moving into the job of fire chief.

5. If you are married, conduct a discussion with your spouse regarding the consequences of making the leap into the chief's job.

COMMITMENT

DAVE HUBERT 9D

CHAPTER 5

Why Do You Really Want
to Become Chief?

Personal Commitment to
the Process

The objective of this chapter is to look at the personal reasons an individual has for wanting to become a fire chief. This chapter builds the bridge between being capable of becoming the fire chief and being the kind of person who will succeed as a fire chief by examining the principles of stewardship and motivation that accompany the pursuit of the job.

Aim high in your career, but stay humble in your heart.
—*Korean folk saying*

THE PRINCIPLES OF STEWARDSHIP

What are your basic reasons for wanting to be the fire chief? I have often said that one of the greatest deficiencies of being promoted to fire chief is that there are no operating instructions on the back of the badge. To complicate things, there is a tremendous amount of technical information for which a fire chief is held accountable. Then there is that mystifying and almost undefinable field of human nature that results in a lot of the problems with which fire chiefs must wrestle. If this job were easy, anyone would take it. But it is not easy. It can sometimes be a terrible burden when you lose victims, especially one of your own personnel, in fires and other emergencies. There are times when people glorify the chief, and times when they vilify the chief. You have to be ready to accept both.

With all of these complexities, what beliefs do you actually have to help you control your behavior and to give you guidance on how to rise above these complications? In one conversation with a fellow chief officer, we reflected on the fact that the more complicated things become, the more we had to simplify our steering mechanisms. My solution to dealing with these complexities is to talk about what I call the basic principles of stewardship.

Over the last couple of decades, I have had the opportunity to discuss principles, policies, and practices with many chiefs from both small volunteer departments and metropolitan departments in this country and on an international basis as well. I have been struck by the fact that I see a great deal more similarities among effective chief officers than dissimilarities. Moreover, the similarities have nothing to do with the scope of their practice but, rather, with the perspective they bring to their job as fire chief.

As a result of these discussions, I have defined what I consider to be ten basic principles. I do not want to imply that these are the only principles upon which to base one's role. They obviously have not been carved on a stone tablet, nor were they passed down to me from a burning bush. In most cases, they have been extracted from experiences of trying to get a handle on complicated sets of circumstances.

The Ten Principles of Stewardship

1. Do not try to know everything, but make sure you know why you are doing anything. Common sense is not a common virtue.

2. Value your personnel as resources and actively seek their input in your decision-making process. Be prepared to decide what you want and what you do *not* want. Further, be prepared to ask for what you want from the people who can give it to you. Value their perspectives, opinions, and observations; do not make critical decisions until you have the appropriate facts. Tell the same people what you do not want. Encourage people who give you what you want by expressing your appreciation to them when they give it you, not later. Discourage people from giving you what you do not want by telling them that you are dissatisfied when they give it to you, not later.

3. Never focus on winning or losing battles but, rather, on the overall needs of the organization. Survival of the principles of good fire protection override individual setbacks. When things go wrong, do not get upset at individuals. Evaluate what went wrong and take steps to ensure that it does not happen again. Remember that all situations are temporary, but relationships last for a long time. Solve situations by relying on facts. Solve problems by building on relationships.

4. Never fear failure; be more concerned about the consequences of failing to try. Seize every opportunity to expand the role of fire protection in the context of your authority. Do not focus on perfection at the expense of performance. Do not listen to people who tell you "it can't be done"; instead, hang around with people who are actually doing things. Never accept mediocrity as a minimum standard.

5. Never promise anything you cannot deliver, and never threaten to do something you are unwilling to ultimately fulfill. Be true to your word. Make your words reflect your behavior and your behavior reflect your words. Take total responsibility for all of your own actions and hold others accountable for theirs. Remember that you can never obtain anything unless you are persistent in pursuing it yourself.

6. Never lie to your boss, your staff, or your subordinates. Keep your integrity even if it costs you a temporary setback. Your integrity is worth more than any decision.

7. Keep the communications between you and your superiors totally confidential. Unless you are directed or authorized to discuss them with others, never violate another officer's trust.

8. Always be fair but firmly committed to act on behalf of the organization rather than on behalf of individual interests. You are ultimately responsible for everyone in the organization. Voting as a decision-making process compromises your ability to take responsibility for the final decision.

9. Always keep your focus on the future rather than on the present or past. Take action today to create your future instead of merely waiting for events to occur.

10. Be realistic but optimistic and committed to succeeding at whatever you want to do with the organization.

That is my list. These principles may not be quite as altruistic as some of the commandments and principles by which we live our lives in the areas of religion and politics, but they have been a fairly good template to apply to my overall behavior and functioning as a fire chief. You need to have a similar set of guidelines, especially before you are given the task of being temporarily in charge of an organization. And I do mean temporarily. Unless you plan on living forever, or never using some of your retirement money, you must admit that you are not the department; you are merely its caretaker. All fire chiefs are stewards of the organization.

There are anecdotes to support practically all of these principles as I have witnessed them in action. They are not necessarily in sequence, but the following anecdotes might be useful in disclosing why these principles can sometimes rise above specifics.

We live in an extremely complicated society. The first principle simply states that the fire chief cannot know everything about that for which he is responsible. I have seen individuals who got themselves into trouble because they were so confident that they were correct on a technical issue when, in fact, the technical answer had changed in recent years, and the previous solution had been rendered obsolete. It is all right to not know everything and still be the chief. What is not all right is to try using old solutions for new problems.

On the other hand, fire chiefs should make sure that they know why they are doing things and should rely on their staffs to know how to get them done. Probably one of the best examples in this area is a situation that had to do with fire codes and changing technical specifications in a fire department. A particular individual who had served as a fire marshal subscribed to the theory that he knew everything the fire marshal knew. When a set of code revisions was introduced at the state level relating to plan check processes, the fire chief chose to move forward with code adoption at the local level before the revisions were adopted by the state agency. In this case, the chief's fire marshal recommended against the amendments because there were secondary effects that were counterproductive in the community. Overriding the concerns of the fire marshal, the chief went ahead and pursued the code amendments and found himself in a compromising situation. When the state amendments failed to pass, he was in the middle of his local adoption process. The question that the chief should have been asking was not when the amendments would be passed, but why they were being adopted in the first place, and whether they would solve his community's problems. Were they absolutely essential to maintain the integrity of a plan-checking process?

A study of high-quality leaders clearly illustrates the fact that most of them were successful in galvanizing the efforts of large numbers of people besides themselves. That is what the second principle deals with, valuing all members of the team, not just the stars. The only way you can gain value from people is to place equal value in all of them.

A story is told at Notre Dame about the so-called four horsemen. They began to believe that they were the entire Notre Dame football team. Coach Knute Rockne sent the team on the field one day with instructions that every time the ball was snapped, the entire defensive line was to lie down and not block the opposition. The highly regarded back field suddenly started getting clobbered. In play after play, the four horsemen were losing ground. They could not understand what was going on. Calling a time out, Coach Rockne brought them to the sidelines and pointed out to the four people whose job it was to carry the ball that they were absolutely useless on a football field unless everyone else was blocking. In the next couple of plays, the defensive line, highly motivated to prove their net worth, forged ahead and took the four horsemen across the goal line.

The worst thing we can do as fire chiefs is to regard individuals as anonymous contributors to the organization. This does not mean that we need to look upon each member of the organization as a prima donnas. We cannot and should not use preferential treatment with individuals. Instead, this principle says we should value everyone in our organization from the brand-new recruit all the way up to the senior member of the department. They will not all be equal in their contributions, nor will they all appreciate the respect that is given to them. On the other hand, evidence supports the fact that the most successful individuals are those who surround themselves with people who are successful in their own right in some way or another. If you value your people, they will value you, and everyone will benefit.

Once again, this has its anecdotal evidence in the fire service, in the experience of a student who took my class on pesticide fire and spill control as part of a state fire training program. When we discussed the use of pesticide identification numbers that are registered with the EPA, this student dutifully took notes. A couple of weeks later, he happened to respond to an incident at a warehouse as a driver's aide. His battalion chief, a seasoned veteran, had fought many fires in his career, but no pesticide fires. Several firefighters got sick. Medical aid was called for, and the individuals were trundled away into the back of an ambulance with specific directions to report to the nearest receiving center.

The problem had a high potential of going from bad to worse. However, the student firefighter remembered the EPA registration number and its relationship to diagnosis, antidote, and medical treatment. He approached the battalion chief, said he had information that the chief might be able to use, and made a suggestion. This was a critical point; the chief officer could choose to value what the student said, or he could discount it. He chose to listen carefully. The firefighters being rushed off to the hospital were given the additional information, and the event had a minimum impact on those who suffered the exposure. The battalion chief was quickly given recognition for effective fire scene management, and the student firefighter eventually rose up through the ranks to assume a higher level of command.

In this particular case, the principles of being in command were not compromised, because the chief valued his people and paid attention. The organizational structure of the fire service, which is based on a theory of managing emergencies that demands a certain degree of conformance, does not preclude us from listening carefully to the input of individuals and respecting their opinions.

There are anecdotes available to support all ten principles of stewardship, but this is only one chapter in a book. The stories provided, however, illustrate the importance of establishing your own set of guiding principles to serve as a backdrop for your decision-making process as a chief officer. As stated earlier, the more complicated our lives become, the more we need a good solid foundation of basics, one of which is a set of fundamental principles on which to base our actions. Life is not simple, but simple principles can render a complicated situation more understandable.

This analogy was brought home to me once again while walking through a museum looking at a collection of exotic carpets. The carpets were quite colorful, with extremely complicated patterns and iterations of symbols that seemed beyond comprehension. Each of these finely made carpets had been made by hand. While reviewing the museum literature, I was intrigued by the fact that the entire explanation for these complicated textile patterns was based on the concept of warp and weave, that is, the longitudinal and latitudinal alignment of thread patterns that was the background for these complicated tapestries.

If we create a warp and weave pattern in our thinking and our approach to the task of managing and leading organizations, we can develop some complicated patterns of organizational behavior. The structure of the foundation will provide consistency and commitment to action. The individual patterns used to fill in the blank spots will give the organization texture and value.

If someone asks you what you *believe* in, can you answer that question in ten statements or less? If you can, it is likely that you have already established the foundation for your behavior. If you have to stop and think about the question, then your response might vary over a period of time.

Pick some principles, and make them your own. Live by that set of principles, and base your career upon them.

FROM BRIDESMAID TO BRIDE: TYING THE KNOT

If you are reading this book, you are probably one of three types of people: a fire chief, a person who aspires to be a fire chief, or a person with the potential for becoming a fire chief who is undecided about whether to enter that arena.

Over the years, I have had discussions with fellow fire officers in all three categories about this job. The conversations with individuals who aspire to become a fire chief and who are looking for some advice on making the transition to chief of the department have been the most interesting. One of the most cogent observations I can make about the fire chief's job is that aspiring to it is nothing when compared with doing it.

There are a large number of people who feel that they could do a better job if they were the chief. Visit almost any fire station around the country, and you will hear a lot of observations about how things ought to be done in the chief's office. Every once in awhile, someone will take his concern for the management direction of the department and aspire to that top job himself. It is unfortunate, but most of the training and educational systems we have in the fire service are not directed toward preparing someone for that position. To the contrary, most of what you learn about being a fire chief is learned after the badge is pinned on your chest.

Nonetheless, there are courageous individuals out there who want to make the commitment to become chief. This chapter is aimed at those who are about ready to make that transition. Those who are already chiefs and have moved from department to department would have other observations to share on this topic, I am certain. The focus in this chapter is on the reasons you might want to move from the position of relative security as a member of a fire force to the vulnerability of the front office.

My observations over the years have resulted in three basic tenets of commitment for fire chiefs. The first tenet is that no one should ever aspire to become the fire chief solely for economic reasons. There is simply not enough monetary difference between being one of the troops and being the chief to make it all worthwhile, especially with the current overtime provisions. Second, one must have a realistic appraisal of oneself, the environment, and one's possibilities of success before making the commitment to become a fire chief, or one will find oneself in a state of conflict, confusion, and frustration.

The third tenet involves the fact that fire chiefs are not really tested for the position; they are selected for the position. The fact that you are technically competent to be the chief is *not* a good reason in itself for aspiring to the position. The unique combination of knowledge, skills, and abilities is what makes an individual a potential fire chief candidate; but the actual selection of a fire chief is based on a combination of those assets and the needs, expectations, and problems that must be addressed by the party selecting the candidate.

REASONS FOR ASPIRATION

One of the main points I make with upwardly mobile candidates is that they must know the reason they want the job before they take it. This may sound simple, but it is not. For instance, I have talked to fire chiefs who have taken the job merely because it was another step on the ladder. Their expectation was that the move to the top job would just mean more of what they had experienced as a battalion or division chief. To their amazement, this was not the case. Others have indicated to me that they competed just because it was an opportunity to be tested against their peers. Often this phenomenon is glossed over as a natural part of the process we use in selecting fire chiefs. Some really want the job, some just compete for the fun of it, and some do not have a clue about why they are there. The real tragedy is when someone in the latter categories gets the job and then has to figure out why he wants it.

There are probably five reasons why a person would try for the job that are based upon personal goals:

Economic gain
Increased power
Increased influence
Increased responsibility
Ego satisfaction

Being a fire chief does have its economic rewards. One should expect that as one achieves a higher level of responsibility in an organization, there should be financial rewards to accompany it. However, the financial rewards that relate to the fire chief's position are disproportionate to the type of responsibility that you are expected to shoulder. Personnel who work in other levels in the hierarchy of the fire department, in many cases, make more money than the chief by virtue of overtime and other forms of compensation. And, in many cases, they can leave their job at the office, at the station, and not have to be concerned about it until they return to their locker. This is not the case with the fire chief.

To the contrary, the job of chief often demands long hours and limited compensation for them. Council and district meetings can consume large amounts of discretionary time. Often the job requires that you engage in expenditures that do not qualify for reimbursement.

Granted, actual financial benefits are there in the form of higher salary for calculation of retirement and add-on benefits such as vehicles and discretionary benefits. These can be offsets to the actual salary paid to a chief. The point here is that becoming the chief is not a windfall for the bank account. If one considers the short- and long-range implications of the economic situation, the chief's job is a long-range benefit but a short-range disappointment. If you choose to be the chief for money, calculate the differences early.

When it comes to economic compensation, an aspiring chief needs to make a realistic appraisal of the benefits to be derived by taking the job. If an individual is being elevated within his own department to the chief's job, there is one set of factors to consider. If the person is leaving one department to become a chief of another, there are other factors to be considered. The commonality between the two scenarios is the fact that an aspiring chief must have a firm grasp of the economic impact of the decision to become the person behind the main desk.

Increased power is also an implication of being the chief, but as we will discuss in a later chapter, the chief shares power with a lot of other people. Power is a double-edged sword in that it flows not from a position but, rather, from a process. You can have power as the chief, or you can lose it to others. Literally, power is a force that is used to demand action. Lots of people are vying with the chief for power. There is the labor force, and there is the city or district governing body. There is even a power struggle going on in many departments between the chief and the immediate staff. If you choose to become the chief to exercise power, you must be ready to respond to those who are prepared to use their power over you. This is not to say that power is not a strong motivator for a person. The quest for it is what causes a lot of people to pursue everything from elected positions to appointed ones.

What is important here is that a potential fire chief candidate realize that power is not the position. Power is what accumulates to people who have utilized the tools of the office correctly to develop the proper relationships. The badge does not give you power. The position does not give you power. Only performance will give you power. Power strategies depend a great deal upon personal relationships, charisma, and your ability to reward or punish others quickly. Power is a strategy that drives an organization in the direction of a few people's agendas.

This brings us to another reason for wanting the job: the ability to influence things. This may seem to be the same as power, but it is not. Power is energy. Influence is much like steering. A person who seeks the chief's job to influence the direction of the department will have to compete with those who are using power to control the organization. But a person who is trying to influence things uses totally different tactics and strategies. Influence is driven by the use of processes that are much more subtle than power strategies. Influence is more dependent upon interpersonal skills and process skills like goals, objectives, and performance evaluations.

As a result, the person who seeks the chief's job in order to influence the direction of the department has to use a wide variety of tools to achieve the desired result. Influence strategies are longer term than power strategies because they take into consideration many factors other than reward and punishment. Key among the influence factors is the ability to convince others that your direction for the department is in their best interests and to convince the authority having jurisdiction that it is in the interest of the community.

If you want responsibility in the chief's job, you can get it quickly. The responsibility of the fire chief's office literally goes on 24 hours a day, 365 days a year. It is a mantle of responsibility that cannot be taken off by going on a vacation, even if you travel halfway around the world. Note that I am talking about responsibility, not action. In many cases, the chief, through proper delegation of responsibility and training of subordinates, can make sure that emergencies are properly handled in his absence. But the responsibility to make sure the organization functions in that capacity rests continuously on the chief, regardless of his location.

I once talked to an individual who came out number one on a fire chief's examination. This individual was excited about the possibility of taking the reins of another department and asked for some personal advice. I began to question him regarding major issues such as salary, benefits, vehicle compensation, availability of the discretionary power in running the department, and so on, and the individual's responses were somewhat ambiguous and confused. Finally, he admitted that he had concentrated so hard on doing well on the test that he did not really know what he was getting into. He might have gotten the job, but he did not get the picture.

If there is any one area where an aspiring chief can find himself in trouble, it is in taking the chief's job without having a thorough understanding of what is expected and of the conditions of employment. The final trumpet on the badge is a great ego booster, but it weighs more heavily than the first four.

It is fair and reasonable for a person who has been offered a job to enter into a negotiation process to protect his own best interests. For example, some of the questions that an aspiring chief might ask his future employer include the following: What is my base salary? Will I be expected to live within the community? What happens if I am terminated within a certain time frame, say three years? With respect to the fire chief's vehicle, is it strictly for official business, or is it available on a twenty-four hour basis?

Expectations, yours and theirs, are going to drive a lot of events in your first year as fire chief. When a person assumes the position of fire chief, he is automatically put into a relationship with both superiors and subordinates in which some form of action is expected. Depending on why you have taken the job, you will develop a set of actions, a list of things to do, that will result in a response by the department. Depending on what you are looking for and what the organization is looking for, there will be a marriage or conflict. You need to be prepared for that.

In the case of becoming a chief over subordinates who have operated under another regime, the subordinates will expect certain things to happen. In some cases, their expectations will be drastically different than yours. The city manager or senior executive who selects a fire chief does not do so merely to fill a vacancy on a table of organization; that person also has some expectations. Your superiors may expect you to take the department in a new direction, or they may expect you to make some significant contributions to improving specific programs in the organization.

Once again, it is imporant for an individual who is being considered for a fire chief's position to accurately assess his personal strengths and weaknesses, this time in connection with dealing with the expectations of both subordinates and superiors. Individuals have been derailed by failing to do so. A derailed executive is an individual who has all the potential for success but, for one reason or another, runs afoul of some aspect of the organization, which causes him to become less effective. One of the best ways to ensuring that this does not happen to you is to do your homework. One of the questions you must ask yourself is, Why do I want this job? If you know the answer to this question, move right along. If you do not know the answer, take a few minutes to determine it. Then you will be better prepared to answer the next question: Should I take this job?

Once you are offered the position of fire chief, it is almost axiomatic that you not accept it outright upon the initial proposal. Like a bride who receives a proposal and then has a period of time to get adjusted before she marries the groom, a potential fire chief should take some time to look at the proposal before tying the knot. This period of time varies in length in direct proportion to the familiarity the individual has with the situation. For example, someone who is aspiring for a chief's job next door is in a totally different situation than someone who is thinking of moving halfway across the country.

There are a lot of horror stories that can be told about fire chiefs who failed to do their homework. There are cases where individuals moved their families and uprooted their entire lives to go to fire departments in which they were the wrong people for the jobs at that time. This does not mean we should run scared of problems or shirk away from responsibilities of dealing with difficulties in new organizations. To the contrary, I am suggesting that a potential chief take the time to do an in-depth analysis of the situation so that he will have a reasonably good chance of success. An individual who accepts the job without doing his homework then has to engage in a form of catch-up after he has the job. He does not come into the situation well-armed and well-prepared with defenses.

Part of the negotiation process in becoming a chief is to make sure that when you identify a specific problem as chief, you will have adequate authority and support by your superiors in resolving that problem. This is a lot more important than most people realize. There are individuals who believe that the rank of fire chief comes with commensurate authority, but this is not necessarily so. There are many situations in which the fire chief position has been compromised, either by political relationships in the community or by binding legal agreements put into place by previous administrations that cannot be untangled. The better you know these predicaments, the more likely you are to be successful in coping with them. Taking the job while lacking this kind of knowledge is like taking a stroll through a minefield with your eyes closed.

Last, but not least, you must be prepared to deal with the issue of selection versus testing. I have talked to a lot of chief officer candidates who were quite distressed because they repeatedly scored as number two or three on a list and never got the job. On the other hand, I have talked to a lot of city managers with the responsibility of selecting fire chiefs who did not care about the candidate's position on the final test and were willing to select individuals further down on the testing hierarchy.

Individuals who are pursuing the fire chief's position should be constantly alert to matching their particular strengths and weaknesses with the needs of the organization for which they are testing. In a lot of cases, the testing procedures are arbitrary and almost insignificant when it comes to the actions and activities that the chief is expected to perform once appointed. Individuals entering the testing process should be less concerned with whether they are number one on the list, and more concerned with whether they are the best person for the job.

After sitting on probably fifty or sixty different fire chief's exams, I can assure you that the individual traits and characteristics of candidates are as important as their technical knowledge, credentials, and experience. You may have heard the old cliché about how one can have many years of experience, or one year of experience for twenty years in a row. Often, candidates come to the desk proposing themselves for the fire chief's position with absolutely no knowledge of the chief's role in the relationship with his superior, how the budget functions in an organization, the major issues that the community is facing, or an assessment of the needs of the organization.

If you want to be selected for fire chief, match yourself with the situation rather than testing yourself against an artificial standard. Know exactly why you are the best person to take the job.

Finally, do not forget that competition is involved. Competition means pitting oneself against others, to put one's best foot forward. It also means pitting oneself against the opportunity to make sure that one measures up to the job. In some cases, that means competing with oneself. Stretch yourself and test yourself as far as you can to make sure you are going for the position for the right reasons.

The job of fire chief does carry with it a certain amount of social status in our communities. Because only one person can be chief at a time, it is axiomatic that you will pass up others with whom you have personal friendships, as well as professional acquaintanceship. In some cases, you will find yourself in predicaments that test your philosophy as well as your technical knowledge. In this chapter, I have used the analogy of courtship because the fire chief's job does involve a relationship. It is a relationship between an individual and an organization; it is a relationship between a person and other individuals. Not unlike romantic relationships between people, this relationship can be ruptured, resulting in estrangement and divorce. The selection of a fire chief has many of the same connotations as the selection of a spouse. The selection is based on mutual self-interests, and it is anticipated that the relationship will last for the duration of their lives. The courtship for a fire chief position should be as realistic and as untarnished by the blinding light of the appointment as possible.

In the final analysis, the best-qualified candidate for fire chief is one who wants to be the chief for a combination of the five reasons discussed in this chapter. The key is that no one factor predominates. When one factor outweighs others, there is potential for frustration and even outright conflict. But when a person seeks the job with a sense of balance between economic rewards and use of power and influence, and with a thorough understanding of responsibilities and willingness to be accountable for them, there cannot help but be some ego gratification. Balance maintains equilibrium.

SELECTED READER ACTIVITIES

1. Develop your own set of stewardship principles.
2. When contemplating a specific job, ask yourself why you think this is a good place to compete. Be specific.
3. Develop a personal statement regarding the reasons that you would like to become the fire chief.
4. Interview a fire chief whom you consider to be successful. Ask him to describe what he likes and dislikes the most about being the chief.
5. Compare the stewardship principles and the reasons you want to become the chief with the list of likes and dislikes from the interview with a chief. Are there items on your list that might result in potential conflict in the future? How do the assets you found in your self-assessment match the positive sides of the job of being a chief?

LEARNING TO COMMUNICATE

CHAPTER 6

Are You Ready for the Spotlight, or Are You a Deer in the Headlights?

Developing Specific Job Skills

The objective of this chapter is to discuss a process that a candidate should use in order to be prepared to compete for a generic fire chief's job, that is, the development of oral and written communication skills.

What we have here is a failure to communicate.
—Line spoken in movie Cool Hand Luke[1]

RISING TO THE OCCASION

By now you are probably getting the idea that the job of being fire chief is tough. It may get tougher. As a matter of fact, a curious phenomenon has

---

[1] Actor Strother Martin, in *Cool Hand Luke,* Jalem Productions, Inc., Warner Bros.–Seven Arts, Inc., 1967.

occurred over the last few years. Many fire departments are actually having a difficult time finding enough candidates to enter into the selection process for fire chief. After all, a person has to give up a lot to be the chief. Overtime is nonexistent for the chief. There may be a loss of personal freedom and a lack of privacy. Sometimes there are a lot of other negatives for the person who fills this job. And those filling the chief jobs in recent years have sometimes been put to the ultimate test of survival well before they had a chance to be taught all of the lessons. The so-called honeymoon period is now down to days or weeks instead of months.

I have spoken with many fellow fire chiefs about situations they have faced, such as forced retirement, their own abrupt termination, votes of no confidence, and conflicts with superiors, peers, and subordinates. Many of them are angry and frustrated. But, to some degree, many of these problems are predictable. We in the fire service, at the chief officer level, are beginning to face what management personnel in the private sector have faced in the last couple of decades: the issue of executive vulnerability. Terms such as *downsizing, executive mobility,* and *hostile takeover* have been part of the vocabulary of the business world for a long time. These terms come wrapped with different implications when they are applied to public sector services such as the fire service. *Downsizing* means reduction in force. *Executive mobility* means that a person can be fired or asked to leave and have to find another job of comparable worth in order to remain in the profession. *Hostile takeover* means consolidation.

One of the consequences fire chiefs are facing for the failure to understand the potential for such events is that they are paying for that failure personally. Many of the job security factors that exist for fire service personnel at the lower levels simply do not exist for fire chiefs. Fire chiefs operate in a different atmosphere.

Although fire chiefs are usually rewarded for their performance, they are also vulnerable as a result of their failures. As discussed earlier, an individual's expectations of the job may not be commensurate with the realities of the job; aspiring to be fire chief is not the same as performing as one. Many dedicated individuals who have derailed once they were promoted failed to prepare for an important aspect of the job because it was not part of the training and educational curriculum: You have to be able to speak and write effectively.

Frequently a new fire chief is expected to make major changes. These changes may be with respect to the performance of a specific person or program, or they may be in regard to individual productivity in the organization. They can be changes with respect to the community and involve new responsibilities or expanded programs for the entire fire agency. You may be required to speak to these issues or prepare written documentation on them.

If you accept the fact that the primary role of the fire service is preparation, then the fire chief should make preparation in the area of oral and written communication skills the number one priority. An individual fire officer must make constant improvements in order to be prepared to face the adversities that emerge from a changing environment. As stated earlier, sometimes the test is given before the lesson is learned. If, in our day-to-day operations, we fail to look to the future with respect to obstacles and problems, there is a possibility that we will face a test and find that we have no resources with which to cope. The fire ground provides ample evidence of this phenomenon, but it is also true in the office.

What is not well-advertised, or even well-documented, is the variety of "real world" tests given to the fire chief sitting behind the desk. I am talking about things like dealing with ethical considerations in code enforcement, serving in an advocacy role when proposing a controversial ordinance, and dealing with a personnel problem in which personal conflict and political forces are brought to bear. In the aftermath of a mass disaster or major operations, the determination of liability, or the finding of fault, seems to be more important, in the final analysis, than the actual command decisions that were made under stress to save the rest of the community.

The ultimate test of anyone in a position of leadership is survival. The adaptation process that we use to survive may consist of power and authority, or maneuverability and intelligence. Power and authority are used almost exclusively in circumstances in which winning or losing is critical. Cooperation and intelligence gathering from a network can result in solutions that are nonthreatening, and this tactic often produces two sets of winners. Verbal and written skills are important to the development of both types of adaptation processes.

The ultimate price that is paid by a fire chief for lack of preparation is vulnerability to attack. It is irrelevant whether the attack comes from sources that are external to the fire department, such as citizens' groups or an irate individual, from an internal source within the municipal government structure, or from within the ranks of the personnel of the fire department. If a person does not survive, the cause is not as important as the effect. The defense to attacks from all of these areas is essentially the same: the ability to communicate effectively. The development of communication skills is a much better weapon than power and authority when it comes to defending against unexpected obstacles and confrontations.

The term *defense* sometimes has a negative connotation, but what I am talking about in the context of this chapter is the ability to rise to a specific occasion. In order to survive a crisis, you must be able to present yourself as the individual who is still in command and control of the organization in spite of the attack. The best insurance policy we can have for competency is a combination of practice and knowledge in communication skills. There is a big difference in the survival rate of individuals who have prepared themselves and the survival rate of those who have merely waited until an opportunity presented itself to demonstrate their courage, integrity, or dedication.

It is not so much a defense as it is a combination of offense and defense—a strategy. The best way of determining your personal perspective on this issue is to look at yourself in the mirror each morning, prior to putting on your suit or uniform or whatever attire is appropriate for going to work, and to ask yourself: Am I better, in some way, today than I was yesterday? Do I have everything I need in my repertoire of knowledge and skills to allow me to provide the leadership to my organization in spite of adversity and complexity? Can I communicate that vision with confidence?

If the answer to these questions is yes, and you are not relying strictly upon a self-fulfilling prophecy that centers around the power of your position, then you may be developing the ability to rise to any occasion. If the answer is no, and you know that you are are still participating in your vocation utilizing knowledge, skills, and abilities that were successful in getting you the job but which have not changed in some period of time, then you had best make sure that when you go to work, you are ready to pay the consequences of failure.

Why risk it? Prepare for the worst of times before you get the job by developing a strategy with which to cope, and the chances are that you will survive anything that can be thrown at you.

THE BIGGEST LITTLE BOOKS YOU CAN BUY

If someone told you that you would be starting your career all over again and that you would be allowed to own only five books for your research, which five books would you choose to put on the library shelf behind your desk? If you were required to manage your organization and utilize only five reference sources to operate your fire department, what five titles would you buy to make sure that you had an adequate amount of information to ensure your own efficiency?

Most of us, when we first started our fire service careers, were introduced to a wide variety of textbooks that are in the field of fire fighting, including such classic titles as the *Fire Chief's Handbook, The National Fire Protection Association's Handbook,* the *IFSTA Fire Protection Publication* series, and, of course, our own fire department's basic training manual, which was almost like a bible when we were going through recruit academy. I am going to assume that you have already read these sources from cover to cover.

Now you are going for a fire chief's job. How many of those textbooks that got you started in your career are really meaningful when it comes to dealing with the problems of long-range planning and interpersonal and organizational dynamics, the realities of obtaining goals, and the prioritizing and assessing of your day-to-day operations? There is a caveat regarding the differences between a person at the bottom of an organization and one at the top: the higher you go in the organizational hierarchy, the less that technical information is really part of your daily operations. The more responsibility you have, the more that the conceptual and interpersonal aspects take on a greater meaning. This applies to the fire service as much as it does to any other industry.

For that reason, I suggest that if you are going to have a library to support your position as fire chief, you may wish to acquire certain books right now. These books have probably never been mentioned in the context of a fire protection administration course or in a fire academy. Nonetheless, these books can have a big impact on your personal skills and abilities if you take them to heart:

Please Understand Me by David W. Keirsey and Marilyn Bates
The Art of War by Sun Yat Tzu, translation by Thomas Cleary
The Elements of Style by William Strunk and E. B. White
The Elements of Grammar by Margaret Shertzer
The Elements of Editing by Arthur Plotnick

Note that the word *fire* does not appear in any of these titles. Yet, each of them has the potential of making a contribution to your effectiveness as a fire officer.

Let us start with *Please Understand Me*.[2] This book by David W. Keirsey and Marilyn Bates is a paperback that deals with the fundamentals of interpersonal behavior. It is a simplification of a contemporary theory regarding personality based on the use of the Myers-Briggs type indicator, which is based on the work of the famous psychologist Carl Jung.

The reason for owning this book (or another book based on the Myers-Briggs type indicator) is simple; it allows you to take a close look at yourself and the people with whom you work. It allows you to identify the

---

[2] David W. Keirsey and Marilyn Bates, *Please Understand Me: Character and Temperament Types* (Del Mar, CA: Prometheus Nemesis Book Co., 1984).

characteristics of your personality style and type and determine which of these elements make you an effective leader or manager. It also gives you a technique with which to explore some of the down sides of each of our basic personality types, which can be helpful in preventing our weaknesses from causing us conflicts and disruptions. It also allows you to take a close look at the makeup of those individuals with whom you work. By coming to an understanding of personality, an individual has the potential of opening up a tremendous amount of communication and establishing better working relationships.

What is important to you, as a chief officer, is your ability to get people to do what you want them to do, and to get them to go in the same general direction in which you wish them to go. The concepts that are explained in *Please Understand Me* do not make an individual a psychologist, nor do they provide all sorts of recipe-driven solutions on how to resolve conflicts with other individuals. To the contrary, the book provides a basic formula for examining personality type and then gives some generic descriptions that one can almost always identify with the names and faces of individuals in one's own setting.

The second book is far more subtle and not very well known in fire circles. *The Art of War,*[3] translated by Thomas Cleary, is based on a book that is over two thousand years old. It was a book originally written by a Chinese general to explain the basic art of warfare. It does not talk about tanks and guns, nor does it talk about aircraft and atomic weapons. It talks about thought processes. For example, here are a few quotes from *The Art of War:*

THE ULTIMATE STRATEGY

To win one hundred victories in one hundred battles is not the acme of skill. To subdue the enemy without fighting is the acme of skill.

ON FOLLOWING THROUGH

To win battles and take your objectives, but to fail to exploit these achievements is ominous.

If you have spent any time thinking about winning and losing, you can probably see that Sun Yat Tzu's psalmlike statements about preparing for war have their counterparts in our daily operations. The book itself has undergone

---

[3] Sun Yat Tzu, *The Art of War,* trans. Thomas Cleary (Boston: Shambhala Publishers, 1988).

many interpretations over the last few centuries. Its simplified, straightforward, tactical and strategical admonitions are just as powerful today as they were when they were first written in China two thousand years ago.

One thing I find interesting about *The Art of War* is that law enforcement people have been utilizing it as a reference in criminal justice programs for years—not in the sense of conducting warfare on the criminal element, but in giving police chiefs a perspective on the ground rules by which they must live in order to achieve community and political support for law enforcement programs. Recently Thomas Cleary updated this classic with a revision entitled *The Lost Art of War*.[4] This book is a translation of Sun Tzu Li's contributions to the body of knowledge about tactics and strategy.

Perhaps it is time for us to take a page out of the manual of warfare and realize that a fire chief's job is not always to fight fire. It is also to fight for principles and perspectives that deal with reducing the fire problem and managing fire resources. General Tzu's book on warfare does not provide a set of hard-and-fast rules, but it does provide a perspective on planning for the terrain and environment of waging a war against apathy, fire, or competition over resources.

The third, fourth, and fifth books in the list have to do with writing skills. *The Elements of Style*[5] by Strunk and White is a book that has been used on college campuses for decades. *The Elements of Grammar*[6] and *The Elements of Editing*[7] are newer books in the series. These books do not tell you what to write, but rather what mistakes to avoid if you want to improve your credibility in writing.

The significance of *The Elements of Style, The Elements of Grammar,* and *The Elements of Editing* is that we live in a world of words. Spoken words and written words often are the messengers used by our political machinery to either support, reject, or evaluate concepts. Although it is important for the fire chief to be able to handle the technical aspects of fire ground hydraulics, it is more important that he be able to handle the technical aspects of communication through written reports.

Over the years, I have reviewed many written reports submitted by aspiring chiefs in the assessment laboratory process or in response to extensive

[4] Sun Tzu Li, *The Lost Art of War,* trans. Thomas Cleary (San Francisco: Harper, 1997).
[5] William I. Strunk and E. B. White, *The Elements of Style* (Needham Heights, MA: Allyn & Bacon, 1995).
[6] Margaret Shertzer, *The Elements of Grammar* (New York: Macmillan General Reference, 1996).
[7] Arthur Plotnick, *The Elements of Editing* (New York: Macmillan General Reference, 1997).

questionnaires provided as part of the selection process. I have been amazed at the number of individuals who will simply write off their skills and abilities by stating up front, in an apologetic way, that they are not "good at writing" but hope that the assessors will not grade them down for their inability to communicate in writing.

Granted, the fire chief of today is not supposed to be a journalist. We are not expected to be world-class writers. On the other hand, when we prepare council agenda letters, position papers on specific issues, or written communications regarding the resolution of issues that are facing our fire department, our writing style says a lot about our thinking style. The more individuals work on developing the ability to communicate effectively in written form, the more they will increase their effectiveness beyond the range of the spoken voice.

WORD WARRIOR

Firefighters fight fires. Writers write things. What happens when a firefighter needs to write about something? Do we know how to "right the write way"? You probably think that the phrasing in that last question is wrong, that it should read "write the right way." One can make a case, however, that writing is something we do to get the right things done. But it must be done in the appropriate way, or it will be unrecognizable.

Doing the right thing often means taking positions on issues that are either undesirable in or contradictory to another person's views. In the fire service, we often find ourselves in the position of defending something that we think is right while receiving a considerable amount of criticism from the outside world. We are often dependent upon written documentation to support our position. One of the most important skills to be developed by chief officers is the ability to express themselves in writing so that they can achieve the goals and objectives of their programs. That is what is meant by "righting the write way."

Firefighters are often very poor writers. I am not saying that there are no firefighters who can write. Obviously, some can write or we would not have articles in our professional magazines, and there would not be any books on the shelf by our fire service leaders and gurus. But the mainstream of the fire service has a reputation for not being willing to or capable of expressing our thoughts in writing with any degree of confidence. This reputation has been developed at two levels: among ourselves, and among those who evaluate our performance, that is, city managers and elected officials.

Many chief officers have told me that they find it difficult to write. It is probably true, because the fire service does not place a high premium on writing skills as you work your way through the rank hierarchy to chief officer. Look around and ask yourself right now, whom among your peers would you classify as being an effective writer? And, although there are exceptions, there seems to be a general consensus among city managers and elected officials that we are articulate in the sense of speaking to an issue, but that we do not demonstrate a great deal of skill in documenting our thoughts. Written documentation from the fire service is frequently evaluated by this group in preparation for policy discussions about fire service proposals. I have heard much more criticism of our efforts to write from this group than I have heard praise.

Among our peers is the law enforcement community. If I may draw a parallel here, this is where our problem may have started. Law enforcement personnel have a different reputation among the people who evaluate them. While I cannot speak to whether individual police officers lament their dislike of writing as much as firefighters, I do know that law enforcement personnel use documentation to obtain both policy and personal support and that, in a nutshell, they are better at using words to win battles than we are. Individuals learn how to write at the patrol officer level, and they continue to write reports up to and including the day that they become a division commander. In many cases, they actually use their writing skills less once they become chief, since then they are in charge of large numbers of people who can communicate effectively on paper to whom they can delegate those tasks.

The reverse seems to be true in the fire service. As we work our way up to the fire chief's chair, we often find ourselves in a position of having to write documents to support our positions but lacking anyone in the organization who has writing skills. We need to change the basic mind-set of our organization and begin developing writing skills at the company officer level. Granted, some of our officers do get an opportunity to develop this skill while working in fire prevention functions or perhaps in training, but the average fire captain has been cultivated to believe that everything is a checklist or a "fill in the box" and, therefore, fails to communicate effectively when writing papers to express attitudes or philosophies or even feelings.

If we want to become word warriors, we need to start at the bottom. All generals need armies to win a war. If the general can write but no one else can, the general is going into battle as a gladiator, not a general. The place to start is in our recruit academies. During a person's initial orientation to the profession, he should be required to express himself in writing to demonstrate that he has this basic skill. The worst thing we have done to our rookies is to expose them to only six letters: A, B, C, D, T, and F. Multiple-guess and true-false examinations tend to provide students with a form of intellectual roulette. If they know the answer, they immediately get credit, but if they do not know the answer, they can guess and be right for the wrong reason. We are not doing them any favors by rewarding them for their luck at chance. At a minimum, we should require as many fill-in questions as the topic will bear just to extract specific facts and to improve upon recall and the ability to spell.

Upon graduation from the academy, personnel are put in fire stations where the only type of writing they are exposed to consists of putting check marks next to items in lists. We only allow the company officers, and sometimes the apparatus operator, to make entries in journals. Whenever an event, such as an emergency, occurs, we expect the officer to complete the narrative. Most of them use a "sighted sub, sank same" type of succinctness in their fire report narratives. Accepting this limited exposure to writing is a mistake. When a person gets assigned to a company, that person should be tutored and evaluated in the preparation of periodic written evaluations of his own experiences. Call it a journal, or a personal log, but having personnel write down their observations, their experiences, and their opinions as a matter of practice creates written skills.

Because the law enforcement community has its patrol officers performing in the area of written skills from the day they go out in a patrol car, they are better prepared in that area as they mature and promote. Because we fail to require the use of these skills, our personnel suffer a form of shock when they finally reach a rank where rudimentary writing skills are required. If you have not developed the basic skills, you certainly will not develop them overnight. Our avoidance of writing tasks at the entry level is largely responsible for the reluctance of officers to write when they need to once they are in key staff jobs.

In our training courses, especially at the community college level, we are letting the skill slide also. I remember the reaction I got from one of my classes when I announced a term paper and essay examination process in a management course. I was heavily lobbied to go for the multiple-guess and "project" approach that minimizes the use of written skills. I did not budge on the issue. It may have cost me a few friends, but it made a few people more competitive as they grew in their jobs.

Another angle to this approach is to use writing exercises in training sessions—on-the-job training exercises. For example, when a person is first learning about fire investigation, an excellent exercise is to require that person to write a comprehensive account of a fire origin, spread, and ultimate destruction. One of my friends, a long-time fire investigator, told me a story that indicates that this was once a common practice. He was required to reconstruct a fire that had occurred at the turn of the century. Originally dismayed at the prospect of finding out anything of value, he asked to review some documents that were archived in Washington, D.C. Much to his delight, reading the reports and writing about the fire made him feel almost as if he had been there.

We are under a lot of pressure to produce today, and sitting down to write out an extensive report is time-consuming. But writing is a skill that is developed through practice. We can be productive and respond to a wide variety of problems, but we can also try a little harder to develop the ability to describe our experiences accurately and thoroughly to others in writing.

When firefighters start telling tales of their experiences aloud, their verbal skills never seem to be lacking. Our ability to describe events in colorful and interesting ways to each other is only a few small steps away from describing them by putting a pen to paper. When I write my columns for *Fire Chief* magazine, I frequently dictate them, a technique that reflects the verbal range of the fire service vocabulary. The point here is that if we can talk effectively about a subject, we should be able to write about it as well.

The next step in the process is to look at the skills we develop as we promote upward in the fire service. Regardless of the level at which a person is performing in an organization, that person can sharpen his writing skills. I have two suggestions for you here: first, take personal responsibility for your own opportunities to practice the skill as you go upward, and second, resist as strongly as you can the urge to do your subordinates' work once you are competent.

In regard to the former suggestion, seek out opportunities to prepare letters, memos, position papers, reports, training manuals, proposals, and even items as technical as specifications or instruction sheets. Seek out these opportunities vigorously; do not wait for someone to tell you to write. Force yourself to try. You are likely to experience problems. Maybe you will even fail. So what? That is what the editing process is all about; that is why first attempts are called drafts. Practice every chance you get, and you will get better and better. At some point, you will begin to think that you are good, and even then you can get better.

Do not let your success go to your head. If you do, the first time you turn the task of writing a paper over to a subordinate, you will be tempted to do the job yourself. That is a trap. It is all right to mentor others, and even to help with the editing, but never take the task away from a subordinate who has been given the job of developing a final draft. When you do for others what they can do for themselves, you steal their capability to grow.

When I was taking my college courses, I had a writing teacher who kept a sign above his desk that read: Write, damn you; write. Becoming a word warrior is a discipline. You cannot write once a year and expect competency. You cannot just talk about it. You have to do it.

We need to place a greater emphasis upon this still in order to increase our professional development. Individuals who are not skilled in writing, or who do not create the skill in others they supervise, often experience difficulties and failures as they move up the food chain. Individuals who have developed the skill, or who at least inspire it in others, are often very successful in getting what they want.

I classify basic writing skills into three simplified steps: getting organized, developing a draft, and editing for readability. These three steps are more simple than most of us believe. The organizational step consists essentially of two minor steps. The first of these is deciding what you have to write about, that is, the goal of the document, and the second step is developing an outline that gathers all the facts pertinent to the case.

Let us say that someone has to develop a paper for the city council explaining the fire department's position on fire safety access for a new subdivision. The fire chief could merely write a short memo to the mayor stating, "We don't agree this housing tract should be built the way it is." Such a memo, however, does not allow anyone to become informed or begin to agree with his position. If the chief clarified the goal and decided that the objective of the document was to inform and gain support from the city council and citizens at large for adequate fire protection in the new housing tract, the context of what the chief was doing would be changed.

Once you have decided on your goal, you can begin to collect all the facts and develop an outline. One of my favorite techniques to accomplish this is to use 3×5 cards. When I am developing extensive documents, I will frequently sit with other individuals and brainstorm the pertinent issues, focusing on which facts are critical and which are miscellaneous. I then arrange them in some sort of logical order, as though I were about to explain the issue to someone who is totally uninformed about it.

After the outline has been developed, the second step is to prepare the initial draft. If one has developed a relatively comprehensive outline with major headings and a number of facts, one can use the headings as major divisions of the report and each of the specific facts to form the basis for a paragraph explaining what needs to be understood.

Most people labor over the draft process by starting at the middle of the process, staring at a blank piece of paper, and exclaiming that they have a form of writer's block. In my opinion, anyone who can talk about a subject ought to be able to write about it. Most of us can carry on reasonably intelligent conversations about issues of which we are to argue the point. It follows then that instead of starting off with a blank piece of paper, you need to write down your thoughts about the things with which you feel most comfortable. This may mean arranging and rearranging paragraphs in the context of a draft document with an eye for the flow of the information. You do not need to write a document by starting with the first word of the first paragraph and going to the last word and period at the end of the last paragraph. The draft can be a "cut and paste" version of your thoughts as you formulate your ideas, perspectives, and what you want expressed in the document.

This leads us to the third phase, the edit. One of the best ways to edit a document is to literally read it aloud. It is not necessary that you have an audience. Reading aloud allows you to listen to the flow of your own words as if you were explaining to people what they need to know. This process may not work extremely well with highly technical information, since a conversational tone may be too casual, but most people feel comfortable when they are reading something that fits their pattern of listening.

When I say edit, I mean edit ruthlessly. Once something has come out of the typewriter, sit down with a red pencil and add and subtract information so it will have its greatest impact. Editing does not necessarily mean cutting whole segments out of the document to make it shorter; rather, editing is the process of making a document more readable and more understandable.

One technique you can use in editing a document is to ask someone who is informed on the topic to read the document and give you their observations. A second party can often look at the words you have framed and give you a whole new perspective on whether the document is succeeding or failing in the communication process. Your worst enemy is someone who gives you back a document with the exclamation "Looks good to me!" You need someone to take an honest second look at your document, and the more important a document is, the more likely it is that you will want to seek additional sources of review.

People who think every document they write has to be letter perfect as it comes out of the typewriter the first time do not understand the power of editing. Recently I watched a program on public television in which the Constitution of the United States was recreated. The anguish that some of the greatest minds in the history of our country went through to develop a document that we all take for granted today was clearly described. Yet, think of how much impact that document has had on the political philosophy and thought processes of an entire world because it was written the right way.

Although most of us fall short of being a Thomas Jefferson or Benjamin Franklin, none of us are lacking in the understanding of our profession, nor do most of us fail to have some point of view that we would like to express or see prevail in making sure things are done the right way in our cities, counties, states, and even our nation. The next time you are tempted to claim that you do not know how to write, say to yourself, "But I do know how to think, and if I can think, I can write."

MAKING A PITCH AND MAKING IT STICK

There are very few natural speakers. Almost everyone, at least once in his life has gone through the terror of being asked to stand up in front of a group of people and make a presentation. Some avoid that experience all the way through adolescence and into adult life. Others overcome the sweaty palms, butterflies in the stomach, and other real and imagined pains of public speaking when they are very young. Some avoid the opportunity throughout their entire lives.

Seldom does the chief officer have the luxury of avoiding that situation. More and more, the fire officer is expected to be able to make a presentation in front of groups ranging from the city council to the Kiwanis Club. Sometimes the chief even has to defend himself in front of public groups regarding a decision that he has made.

As a matter of fact, one of the dimensions that is frequently examined in the assessment laboratory process to determine future career opportunities is called oral communications. Many people think that this is a test of being able to give a training presentation. It is not. It is a reflection of the need to be able to defend yourself verbally. So, it behooves every fire officer to develop public speaking skills.

The problem is that most fire officers do not have the opportunity to practice this skill until it is time to test for the next promotion. Many individuals who serve as key staff officers without having to make public presentations find themselves uncomfortable when they are appointed to the fire chief position and have no one else to make presentations for them. Waiting until the last moment to develop public speaking skills is almost a blueprint for failure.

I have had an opportunity to observe a large number of chief officer candidates who have been asked to go through the assessment exercise of conducting a verbal presentation. Unfortunately, I have seen many very qualified candidates literally melt under the stress of this one exercise. It does not have to be that way. One does not have to be a highly polished speaker in order to make an effective verbal presentation. Public speaking is a skill that can be learned just like any other skill. One does not even have to enjoy public speaking in order to be effective. There are many individuals who are recognized as effective public speakers who are highly introverted and engage in minimum conversation on a one-to-one basis. But they have mastered some of the basic skills of public presentations, and they receive recognition and response from their audiences.

There are many good textbooks and courses available on public speaking, and it is not my intent to repeat the more technical aspects of it. Instead, this section focuses on the triangle of presentation that has only three considerations. The first leg is preparation, the second leg is planning, and the third leg is poise.

Regardless of the nature of the speaking engagement, these three elements must be dealt with in sequence if an individual's presentation is to be effective. Contrary to popular opinion, some of the most extemporaneous speakers undergo the preparation step. Planning a presentation is every bit as important as delivering it, for without a plan, any speaking engagement can take on the aura of a rambling monologue. Poise is not the same as charm. Poise is the ability to remain physically under control despite the stresses created by the speaking engagement. Let us deal with each of these in a specific sequence.

It is rumored that once upon a time, an individual stopped a famous musician outside Carnegie Hall. The individual, who was trying to find a way into the hall to watch a presentation, asked the musician, "Do you know how to get into Carnegie Hall?" The famous musician looked the individual straight in the eye and replied, "Practice, my son, practice!"

There are literally thousands of opportunities to practice public speaking in the context of the fire service right in the firehouse. Conducting a training session for one or two personnel may not be the same as delivering an address to the United Nations, but it is an opportunity to develop oneself as a speaker. The preparation phase of public presentations consists of refining one's comfort factor in communicating with people first on a one-to-one basis, then in small groups, and ultimately to larger groups.

Preparation can take place on many levels. For example, most communities have an organization called Toastmasters. This group provides an excellent environment for even the most nerve-wracked speaker to learn by doing with a minimum of risk. A small investment in a meeting of Toastmasters can reap many benefits down the road.

Modern technology can be brought to bear. For example, an individual who makes a presentation under whatever circumstances can videotape it and personally critique it at some later time. Unfortunately, most speakers have no concept of how they sound or appear when speaking. This lack of familiarity creates anxiety. Even the simple technique of tape-recording a presentation can result in an individual gaining insight into his delivery. Preparation consists of getting yourself comfortable with the position of being the center of a presentation.

During the preparation phase, one should develop a technique for collecting one's thoughts and learn how to utilize it under stress. For example, early in my career, I was shown a technique for organizing a speech by using 3×5 cards. But, more importantly, I was later shown how to punch a hole in the corner of the 3 × 5 cards and put a ring through them so that if the cards were dropped, the accident would not destroy my train of thought.

The preparation phase of public presentation is actually one of self-assessment. It is an examination of yourself, your personality, your communication style, and your own individual strengths and weaknesses. If you examine the types of people who are communicating to us through the media of both television and radio, you will soon discover that there is one of everything. There is no such thing as a perfect speaking voice, there is no one volume that always works, and there are effective communicators in all sizes and all shapes.

As alluded to earlier, one of the most important elements of preparation is feedback. Good honest feedback from someone with whom you can share your feelings is an essential part of the preparation phase. The Toastmasters are experts at this, but you do not have to be part of a formal organization to get the benefit of good critiques.

Preparation leads to the second leg of effective presentations, planning. You would be amazed at the number of people who do not know how to start a speech, how to end a speech, or how to properly structure the middle of a speech. Each part of the speech is equally important. Most effective speeches have only four elements to them. The length of time spent on these elements varies in accordance with audience and environment, but basically the four elements are straightforward and repeatable from presentation to presentation. These elements are represented by the acronym AIDA, which stands for attention, interest, detail, and action.

An effective presentation always begins with some form of getting the audience's attention. If you read a newspaper article, the first paragraph, called "the lead," is usually something enticing to get you to stop scanning the page and pay attention. The same concept applies to a good verbal presentation. Effective speakers provide the listeners with some sort of audible indication that it is time to pay attention.

The second element, called the interest statement, is nothing more than a brief statement that tells the listeners why they should be interested in the topic. The more simply the interest element is stated, the better. Long-winded messages lose people quickly. Most effective presentations tell the listener right up front why the material is going to be important to them personally.

Detail is exactly what it sounds like. It is the organization of the facts that you wish to present to the group in some systematic and logical fashion. There are as many ways to arrange details as there are types of details that can be presented. Some speakers find this phase extremely easy. Others find it difficult. Nonetheless, the detail portion is the meat and potatoes of a verbal presentation. As mentioned earlier, techniques learned during the preparation phase can be extremely valuable in bringing the details into focus during a presentation. The use of prompting devices can serve to structure this portion of any verbal presentation. I have seen speakers use techniques to organize the details of their speeches that cover everything from notes written on the palm of a hand to the use of teleprompters in television studios.

The body of every presentation must contain a sufficient amount of substance to support the final step, which is the action phase, or what salespeople call "the close." In essence, the action element is the end of a verbal presentation where you ask the listener to do something. It may be to pass an ordinance, support the local fire department, make a commitment to buy a smoke detector, go out to vote, or whatever. The most effective closes are action summaries related to the interest element that was stated in the beginning of the presentation.

The third leg of verbal presentations is poise. Another term for poise might be discipline. Believe it or not, poise is something that can be studied and developed. One of the best techniques for improving one's poise is to make a conscious effort to slow down one's speech while, at the same time, doing something effective with one's hands and feet. For some reason, many people have been led to believe that the most effective speakers are those who use a lot of hand gestures, but that is not true. Comfortable hand gestures and body movements are so subtle that very seldom distract from a speech, but they seldom add much to it. A constant movement of hands and feet and the shifting of a body's position in making a verbal presentation are very distracting.

I once observed a highly trained professional making a presentation to another group of highly trained professionals on a very controversial subject. I was involved in working with both groups and knew the speaker on a personal basis. I was also very familiar with the speech and, therefore, was not particularly listening to the content as the individual made his presentation. His calm, deliberate, and well-paced presentation was being delivered to the audience in an effective manner. Then I noticed that, from where I was sitting next to the podium, I could see his knees shaking in a visible fashion. No one else could see the movement, however. The individual was in total control of everything that could be seen by the audience, but the anxiety had to come out somewhere, and it came out in his knees. That is poise.

Later, I asked that individual how he felt when he was making the presentation, and he said, "Scared to death." The audience never knew that. As I recalled his presentation, I remembered that as he made his speech, he stood firmly grasping the podium with one hand, relying on another individual to turn the transparencies from page to page on the overhead projector. He had prepared for the speech, had a plan for the speech, and exercised discipline to control an almost pathological concern over how the material would be handled by the audience. Several months later, I watched the same individual make the same speech to another group. The material had not changed, but his polish and effectiveness in delivering the speech had substantially improved.

One does not have to become an expert speaker in order to do a good job. There are so many people who have a fear of making public presentations that a person who becomes just average at it will leave a favorable impression. With continuous preparation, the development of a plan for each presentation, and a continual evolution in your ability to control your body during speeches, you will not only eliminate the butterflies, but you might be able to soar with the eagles.

SELECTED READER ACTIVITIES

1. Develop your own personal library on writing and speaking skills.
2. Purchase a copy of *The Art of War* and compare its concepts to the problems faced by a fire officer.
3. Seek out every opportunity to write, edit, and review written documentation in the fire service.
4. Seek out every opportunity to make verbal presentations to groups both inside and outside to the fire service. Attempt to diversify the presentations so that some experiences are improvised, and some are planned. Some of the presentations should be advocacy speeches, while others should be informational.
5. Join a local service organization and participate in the processes used to develop its leadership.

BEST FOOT FORWARD

CHAPTER 7

Putting Your Best Foot Forward:
Actually Applying for the Job

The objective of this chapter is to discuss how you apply your inventory of skills to the competitive process. The manner in which you present yourself in writing and in person is going to have more to do with the selection process than you think.

The great difficulty is first to win a reputation; the next to keep it while you live; and the next to preserve it after you die, when affection and interest are over, and nothing but sterling excellence can preserve your name. Never suffer youth to be an excuse for inadequacy, nor age and fame to be an excuse for indolence.

—Benjamin Haydon (1786–1846)

THE RÉSUMÉ

It would be interesting if, when you died and went to heaven, the first question asked by the angels outside heaven's gate was, "Do you have a résumé?" How would you like to have your life evaluated on the basis of what you put in your résumé? We will never get a chance in this world to determine the outcome of that fantasy, but we have something almost as important, that is, using a résumé to apply for the job of fire chief.

Your résumé is you! What you put in that résumé can determine whether you will even be invited for an interview. It may determine whether you will become a finalist for the job. Therefore, you should examine the résumé writing process.

The material in this chapter reflects my personal biases. There may be people who would disagree with some of the guidelines I suggest, but my rationale is based upon almost thirty years of reviewing résumés for job applications. These suggestions and warnings are based on notes I took during selection processes.

Let us deal with the suggestions first. One of the first things you need to do in developing a résumé is to create a résumé *file*. All of us do things in our lives and then forget about them. Our day-to-day experiences are often obscured by the months, weeks, and years that follow. A résumé file is a box, a drawer, or just a place on a bookshelf where you keep track of your life experiences. You want to get as much credit as possible for the things you have done, and you do not want to forget them when you are preparing a résumé.

Résumé writing is not so much a case of describing your name, rank, and serial number as it is painting a picture of yourself as an individual. Therefore, you need to create a résumé file in which you can collect all of the things that best describe you and what you have done with your career. Some of the information is vital. Other items are useful only in context.

The second thing to do is to create a generic résumé. It should be a description of your life in reverse. It should start with where you are today and go backward in time, not unlike a highway that goes off into the distance as we get further and further from today. Things become less easy to visualize as you proceed into the past. Events that occurred twenty-two years ago are not as important as those that occurred two weeks ago. You need to develop a generic résumé, however, that tracks your life's experience from now to way back then.

Third, when you get ready to apply for a job, *review the job specifications* and make an instant comparison with your résumé. It will not make a lot of sense for you to seek a job that requires certain kinds of experience, knowledge, skills, and abilities unless you possess them. To be a successful job seeker, you need to make sure you, as a candidate, are the type of person being sought.

Create a custom résumé for every specific job you seek. People who review résumés can almost smell generic ones. The generic résumés tend to look like autobiographies. A common complaint I have heard in the review process is that a person's résumé is so lengthy, the reviewers cannot find the pertinent facts. Therefore, that résumé goes into the rejection pile. What your résumé should contain is evidence that you have already done, or are prepared to do, that which is being sought in the selection process.

With computer word processing, there is absolutely no excuse for using a generic résumé. You can leave the generic résumé on the screen, hit the "save as" button, and throw away everything that is not pertinent to the position you are seeking.

The last thing to do is very simple. When you submit your résumé, make sure it is crisp, clean, and succinct. *Never* send a photocopy. If a résumé encompasses more than two or three pages, it is probably a case of overkill.

Now let us look at the warnings. Never embellish your own experience. In preparing a résumé, always keep the description of your personal experience accurate. Do not take credit for an overall project in which you actually played a relatively minor role. If you participated as a committee member, simply say you were a committee member. I once had the dubious distinction of an individual appearing before me in an oral board whose résumé included a project in which I was the project manager. I was amazed when I read his description of his role in the committee's activities. I had been there when all the activities transpired and knew his role was limited primarily to a minor portion of them, yet the résumé painted his participation in this project as being the equivalent of my role as project manager.

Next, do not try to hide a negative experience. Sooner or later, anything that has happened to you is going to come out in the background investigation if you are a successful candidate. Therefore, if you have a black mark, even if you have been terminated, just say so. If you have had any negative experience in your career that will come out in a background investigation, lay it on the table. Do not try to obscure it with the use of fancy terminology and vague descriptions. It will come back to haunt you sooner or later.

Do not be too cute. Some people try to make their résumés stand out by using special binding procedures, fancy type fonts or colored paper. They believe that these things will make their résumés stand out in a stack of other résumés. They do. However, they also make them far easier to remove from the pile. A simple, straightforward, well-designed résumé speaks more of your competency than one that appears to have been done by the marketing manager for a fast-food restaurant.

Do not be discouraged if you are not accepted in a résumé screening. Rejection on the basis of a résumé is not rejection of you as a human being. It merely means that whatever they are looking for may not be what you are. The worst thing you can do is become overly sensitive to rejection and start dealing with evaluators as the enemy. They are not your enemy; they are people just like you and me. Their job is to find the right person who meets the job specifications.

SWEATY PALMS; BUTTERFLIES IN THE STOMACH

In the last chapter, we noted that one of the most commonly held fears by most people is that of being asked to make a speech in public. This concern over being asked to stand in front of total strangers and expose oneself to criticism is probably why people also fear oral boards. No one likes to be evaluated by another person, although all of us are evaluated all the time by our families, friends, and even our enemies. What is the basis of the fear? No one wants to look bad.

Many people are open about their fear of taking oral boards. I have seen some of the most courageous firefighters turn pale at the very thought of having to sit at a narrow table and be scrutinized by someone. Some of the most intelligent individuals allow themselves to be dumbstruck by a panel of individuals with lesser IQs.

Yet, there are those who survive. Someone must be surviving this process or we would not be promoting people. As a matter of fact, some people not only survive, they thrive on orals. Often the individual who is least qualified to take the job is better at taking the oral examination than the best candidate. The outcome of that phenomenon has been discussed in many fire departments.

I have also heard it said that many individuals have built their careers upon the ability to "B.S." The oral communication skill is not something that any of us can take for granted. If it is used as a means of assessing qualifications for promotion and as the basis for the credibility of our reputations, then perhaps we need to take a close look at this concept of oral examination and see if we can improve our skill in this area.

There are two sides to this coin. First, there is you as a candidate, you as the individual who is taking the oral examination. And then there is the person on the examiner's side of the table. That could be you also. If you are preparing yourself for the position of fire chief, you should have some experience in both venues. It is not uncommon for a person to be asked to sit as an evaluator on an oral board once that person has achieved a certain level of rank in the fire service. For purposes of this discussion, both sides of this coin will be examined because they are related.

I once sat on an oral board and screened a large number of candidates where I had a lot of personal knowledge regarding their backgrounds. Many of them were people with whom I had worked closely as a committee member or in a professional capacity. The other members of the board were experienced individuals in their own right; however, two of them were not fire professionals. One candidate who came into the room and sat down was visibly shaken. I had known this individual for more than twenty-five years and was startled at his response. All of the classic symptoms were there—sweaty palms, hand wringing, beads of perspiration on the temples and upper lip, darting eyes, and stiff posture. I could not help but feel that the person I was seeing at on the other side of the table was a caricature instead of the experienced officer I had known over the years.

As the questioning continued, it became increasingly clear that this person was more than nervous—he was terrified. On answers that required brief responses, he became wordy. On questions that required some in-depth thinking, he was superficial. When he shook our hands before he left the room, I could feel the aftereffects of his tension and anxiety. He was obviously glad the experience was over.

After this candidate had left and everyone had graded him, I asked the other board members if I could depart from normal procedure for just a moment and ask them a personal question. I prefaced my question with the comment that I did not know the scores on the test, nor did I wish to see them modified, but I was curious about their perceptions of the last candidate. Without belaboring all the points, they basically said they felt he was a very good candidate but was obviously badly distracted and did not give a good oral presentation. I asked them if they had ever had an experience of giving a bad oral presentation themselves in their upward mobility. The evaluators replied in the affirmative by stating that in many cases, they were probably more intimidated by an oral board that consisted of individuals they knew very well.

All of us have survived these crises and this particular individual did so as well. He was allowed to pass the examination, and was later chosen as the best candidate for the job based on a background investigation. His performance demonstrated that he did, indeed, have the capacity to serve, but he had risked a lot by not being better prepared for the oral examination.

The oral board has two elements—the interviewee and the evaluators. The process that goes on between them is one of communication—oral communication. Many people carry the fear that the purpose of an oral board is to strip away their veneer and expose all their weaknesses and failures. Although I cannot argue the fact that some interviewers think that purpose is an appropriate role for the board, the process itself is not designed to do that. The process is designed to give you an opportunity to verbalize your knowledge and to present yourself in a way that demonstrates your skills and abilities to perform the job for which you are testing.

With that in mind, I would like to carry you through an oral board process and describe some things that a person should do if he wishes to be a successful candidate. Granted, some people will be better at using these techniques than others. Some will find the techniques to be nothing more than second nature. I have sat on several hundred oral boards where these observations have been reinforced time and time again. I have seen very few "tens" and very few "zeroes." I have seen individuals with a high degree of potential demonstrate their skills, and I have seen other individuals with very low potential make themselves better merely by applying common sense and paying attention to what is happening.

Do you remember the comedian Flip Wilson? One of his lines was, "What you see is what you get." An oral board is an example of that statement with one slight addition: "What you see and hear is what you get." Some argue that an oral board examination is relatively subjective, and in many ways it is less objective than assessment laboratories, and it is much less technical than paper-and-pencil examinations. Yet, it is also objective in that it measures an additional dimension of a person's skills and abilities. When good candidates go up against good oral boards, the process is almost electrifying, but when poor candidates are matched with poor oral boards, the process is dismal and frustrating and, to one degree or another, humiliating for the participants on both sides of the table.

How do you study for an oral examination? The first thing you have to recognize is that you cannot study for specific questions in an oral board. You have to prepare yourself mentally for being ready to articulate what is already in your mind. Do not memorize statistics, facts, figures, and responses in hopes of regurgitating them in an appropriate sequence for the board; instead; instead, prepare yourself mentally by studying to understand the width and breadth of the job for which you are testing. You may be the world's leading authority on the job you currently fulfill, but the real issue is whether you know anything about the job for which you are competing.

Reading in the topical area and spending time with individuals who can mentor you in the tasks, responsibilities, and activities of that new position are important parts of the preparation phase. However, you also must learn to think like a person in that new position. I recall an individual who was a tremendous asset to my career who taught me that lesson.

Captain Larry May was my captain at fire station number two in Costa Mesa. One day he called me into the office, told me to sit, and asked me a series of questions about the job I was currently filling—fire engineer. He then asked me, "Are you ready to become a captain yet?" When I answered in the affirmative, he pushed his chair back from his desk, stood up, and motioned for me to sit down in his well-worn seat as he said, "Well, have a seat, because I'm going to teach you how to think and act like a captain." When it came time for my oral board, I did not have to talk in a theoretical sense about things like managing company time or conducting training in the fire stations. I had already accomplished these things and was merely reflecting on the experience.

Another part of the preparation phase is to pay very close attention to your own personality. Earlier we talked about understanding the strengths and weaknesses of various personality types. Knowing whether you are an extrovert or introvert, or whether you are an intuitive person or a sensing type of person, will have a big bearing on how you prepare yourself to respond. For example, I often listen for key words when an individual responds to a question. If I ask a technical question and a person responds with, "Well, the way I feel about that is . . .," I know that the person is not using the proper behavior to respond to the question. He is telling me his feelings instead of the facts. On the other hand, if I ask a question that is heavily loaded with organizational or human dynamics and a person responds with, "Well, the way I think about that is . . .," I recognize that the person may or may not be giving proper consideration to the needs of other people.

Understanding the strengths and weaknesses of your own personality type and how they are reflected in your vocabulary and in your personal behavior is an important part of preparing for oral boards. If you are asked a question that requires analysis, you must use the thinking side of your brain. If you are asked a question that deals with human nature or some interpretation of reality, then you must often use your intuitive and feeling skills.

Then there is the matter of attitude adjustment about the oral board members. I have seen people go through a lot of stress and strain to find out the names of the people who will be sitting on the oral board. The basic attitude you should have about evaluators is that they are there to do a job; it should make no difference whether or not you know them. You need to recognize the evaluators as individuals and be responsive to them when they make their inquiries. There are many crazy things that can happen during oral board processes to which you must be ready to respond. I have seen oral examinations where only one of the board members was allowed to speak, and the other board members merely wrote down the responses of the candidates. In other oral boards, all the evaluators not only spoke but also asked additional questions and engaged in an actual dialogue with the candidates.

Then comes that fateful day when your appointed time to appear before the oral board arrives. You are going to have a specific number of minutes to make a first impression. There are a couple of things you might want to do prior to that appointment. Take a look at your personal grooming. It is not a good idea to get a haircut the day before you go to an oral board examination; individuals who do that often look like a peeled onion. It is also not a good idea to make a major change in your style of dress or even your diet. Keep your lifestyle as normal as possible prior to the actual date of the oral examination. If at all possible, get a good night's sleep, and follow your normal dietary practices by not skipping meals or overconsuming alcohol the night before or drinking too much coffee that morning.

Now you are sitting right outside the door. The people running the board may or may not provide you with an opportunity to be reasonably comfortable, but in either case, that should not make any difference. You may see the previous candidate leaving the room. That can create a number of different reactions. Very confident individuals have told me they liked seeing the nervousness and anxiety of their predecessors, and underconfident ones have told me that it bothered them to see a previous candidate stride out of the room with a smile and a gesture indicating that things went well.

You may not have a choice about whether you are in the room when the previous candidate leaves, but you do have some degree of control over your reaction to seeing him. The worst thing you can do is to ask him how the examination went. It is best to smile and be cordial, but not overly curious. If the other person thinks they are a better candidate than you, they may try to make you feel more awkward. A person who thinks they are a lesser candidate than you may try to give you misleading information. You may find this hard to believe, especially when you are competing with friends, but competition brings out some strange behaviors.

Now it is time to walk through the door. Sometimes you are asked to enter by an anonymous voice on the other side of a door, and sometimes someone escorts you. The thought that should be going through your mind as this point is that your primary objective is to leave the oral board with a favorable impression of your skills and abilities as they relate to the job for which you are competing.

From this point on, a lot of things can happen, depending on your ability to control yourself and, to a certain degree, even control the oral board process. That might sound odd, but there are differences in the way in which interviewees handle the oral board itself. A subtle application of time management usually is not only beneficial to you but also appreciated by the board. For example, if you notice that candidates are scheduled about thirty minutes apart, you will need to pay attention to how you answer your questions, since thirty minutes is not much time for three to five board members to ask many questions. On the other hand, if you notice the orals are scheduled one hour apart and there are only two interviewers, there is a possibility for some in-depth questioning.

One of the first things you should do is size up your assessors. This is actually an ongoing process during the entire oral board interview. Initially, pay attention to two things with respect to the oral board: the physical configuration and the board members' physical appearance. Physical configuration has to do with whether they are sitting at a table directly across from you or whether they are sitting in a more informal arrangement. Physical appearance refers to whether the members are dressed formally or informally, and what kinds of writing materials are on the desk in front of them.

Another subtle implication of control is whether you are introduced to the oral board. If the person who brings you into the room does not introduce you, it is not inappropriate for you to offer the basic courtesy of a handshake before you take your seat. If you know someone by name or rank, it is not inappropriate to say, for example, "Nice to see you, Chief," or something of that nature. You should not, however, become chatty and attempt to engage the oral board in a dialogue. Rather, be courteous, and then take your seat.

At this point, several dimensions take over to create an impression on the board. They can be classified into two categories: body language and thought processes. Body language breaks down into posture, eye contact, and physical control. Thought processes break down into vocabulary, humor, and response.

Let us first deal with the physical aspects. A person should be comfortable when taking an oral examination. Being perched on the last three inches of a chair creates a sense of tension in the candidate and in the evaluators. I once sat on an oral board where we (in retrospect) noted that we had actually taking mental notes of a particular candidate's position on the edge of a wheeled chair. This individual had been sitting on the front end of the chair and had moved the chair back about two feet from the table. It was almost as if he was in a state of suspended animation.

Your posture should be relatively comfortable, but you should never allow yourself to slouch or rotate your buttocks in the seat of a chair to give the impression that you have "kicked back." You should place your hands in your lap or on the arms of the chair and use them as you would in a normal conversation. Constant fidgeting with a ring or wristwatch, twirling the bottom of a tie, or tugging at an item of clothing distracts the evaluators from what is actually being said.

Eye contact is part of body language. When someone is talking to you, you should be looking at them. A person who looks around the room and becomes enamored with a light on the ceiling or stares out a window when being questioned by an evaluator or, even worse, when responding to an evaluator leaves the impression of indifference and evasion. It is extremely important that you maintain eye contact with anyone who is questioning you. Equally important, if the question has a lengthy answer, you should exchange eye contact with each member of the board at some point during your response to determine whether each person is listening to what you are saying.

The response aspect has some interesting dimensions. We talked earlier about thought processes. Many people think that when they are talking, people must be listening. But listening is not the entire task of the oral board; the board puts into an appropriate context how your thoughts bear on the type of candidate being sought.

One of the most frequent mistakes I have seen is that of individuals giving the right answer to the wrong question. It is important that you pay close attention to the question being asked. You should answer *that* question, not an interpretation of it or a variation that meets your own personal interests. Be careful to keep in balance the length of your responses. I remember one board where we had been given only thirty minutes to talk to the candidates for rating purposes. One individual who was asked a relatively simple and straightforward question talked for seventeen minutes about that first question and was left with only a couple of minutes for each of the other six questions on our list. You should always try to be as succinct and, at the same time, as comprehensive as possible. As a practical rule, anytime an individual talks more than three or four minutes on any topic, he may well be wearing out his welcome in the ears of the evaluators.

Another frequent mistake is making an attempt to use humor when it is inappropriate. You may be very funny in the firehouse, and your sense of humor may be recognized far and wide by your subordinates, but a sense of humor used in an oral board has to be very carefully balanced with the values of the people across the table. They may not think that any of their questions have the potential to be humorous.

Another mistake to avoid is making any remarks that have ethnic, racial, sexual, religious, or even institutional biases associated with them. Be extremely careful to avoid such remarks. Associated with this is the use of your vocabulary. Be careful not to use a lot of "ah's" or slang terms, such as "yeah" for yes or "you betcha" for yes, I agree.

When taking an oral examination, do not bluff. If you do not know the answer to a question, you are far better off simply stating that you do not know the answer. Not knowing an answer does not mean you are stupid; it merely means that you are not aware of an appropriate response. Evaluators would rather have the person admit not knowing the answer than attempt to wind a way through a convoluted bluffing process.

It is now time for the oral board to be over. It is not uncommon for the final question of the board to be something like "Why do you think you are one of the better candidates for this job?" Many candidates believe it is appropriate to be humble at this point and state, "Aw shucks, I'm just one of the boys trying out for the job and really don't feel I have anything special to offer." If you make such a comment, you might as well take the eraser and reduce your points by about twenty-five percent. You need to remember that you are in a competitive process, and in competition, someone has to come out first, and someone has to come out last. You should always be prepared at the end of every oral board examination to give some kind of accounting of yourself. If you have reasons for wanting the job, you should state them. If you spent a great deal of time preparing for the job and specific experiences indicate that, you should discuss those experiences. Humility is all right for sainthood, but will not get you ready for promotion.

Finally, the oral examination is over and you leave the room. The oral board process, however, should not terminate the minute your hand leaves the doorknob. Instead, go somewhere quiet, take a piece of paper and a pen, and jot down some notes about the experience. Remember the kinds of questions that were asked and your specific responses. You might try to recall how you felt when making the responses. Did you feel confident, or did you have misgivings? You may wish to jot down some memories of what the assessors were doing when you were providing answers. Did they take notes during any particular responses? Did they all take notes, or did just one person take them?

"When all is said and done, more is said than is done" is a good motto for oral boards. An oral board process is an opportunity for you to make a visual and verbal impression upon other individuals about your capabilities. If you have done a good job, you will get a good score. If you have done a poor job, it is possible that you will not be rewarded. Do not blame the oral board.

If you have taken an oral board and have been promoted, congratulations. But do not eliminate them from your experiences. Consider entering phase two, which involves sitting on the other side of the table as an evaluator.

Everything that has been said about the candidate's participation should also be part of the evaluator's considerations. Many evaluators focus on things such as posture, vocabulary, and eye contact. All of these leave impressions; however, there is more to doing a good job of evaluating candidates than merely determining whether they fit a preconceived notion of what a person should look and sound like for a particular position.

From an evaluator's point of view, there are at least two ways of classifying information for purposes of evaluation. The first is measuring specific behaviors, and the second consists of evaluating specific verbal responses. The oral examination is a visual and verbal situation. The members of the oral boards do not know what a candidate is really thinking; they can evaluate only what the candidate is doing or saying.

The first thing an evaluator needs to be concerned about is the job description and the dimensions for successful candidates. The job description contains specific information that candidates should be prepared to demonstrate—knowledge, skills, and abilities for specific functions that are usually slightly different from those the candidate uses in his current position. Therefore, one of the first things a good evaluator will do is to review the job description and the dimensions that have been designed for successful candidates. Since the evaluator will review this material, it is important that the candidate do so as well.

The second step for an evaluator is to carefully examine the scoring mechanism and attempt to determine what specific behaviors or responses would distinguish an average candidate from an above-average candidate. Although the oral examination has been considered subjective, the fact is that specific responses to specific issues can be fairly clearly defined in advance. Suppose a candidate was being questioned about his knowledge of tactics and strategy. An individual who is able to list the various divisions of fire tactics and strategy may be an average candidate. An above-average candidate may be able to not only list them but also provide definitions and a more adequate explanation of the incident command system as it relates to the division of tactics and strategy.

The third phase of evaluator preparation is to look very carefully at the instrument used for scoring the candidate. The form should have some relationship to the dimensions of the job for which the candidate is being evaluated. The best oral examination forms usually provide an area in which the evaluator can make specific notations to support his placement of an individual into a specific category. Forms that contain an overabundance of check boxes and that do not contain definitions of the dimensions to be evaluated place a heavier burden on the evaluator.

The last phase of preparation for an evaluator consists of a discussion with the representatives of the organization about the types of candidates for whom they are looking. An evaluator has to be careful not to misuse this by discussing individual traits of specific people. This could lead to a type of bias that would be counterproductive in the evaluation process. It is reasonable to expect a fire chief or battalion chief to sit down and discuss the needs and expectations of his organization. If the organization expects their officers to think independently, that should be discussed. If the officers are expected to comply strictly with operational policies and procedures, then conformity may be an important dimension. This discussion should focus more on the needs of the organization than on a description of the "perfect candidate."

Then, the candidate comes through the door—the moment of truth for that individual and for you. An oral board evaluator should begin this relationship with professional courtesy and a certain amount of empathy. After all, you have been there before. Establishing good eye contact and participating in the handshaking ceremony, along with a few brief informalities to establish a relationship, can do a great deal to put the candidate at ease and allow the evaluators an opportunity to gain maximum benefit from a short time together.

During the preparation phase, there is often an agreement made among the evaluators to ask specific questions and to ask them in a specific sequence. For purposes of uniformity and standardization, it is always a good idea to ask all the candidates the same questions in the same way. While it does not preclude evaluators from doing follow-up questions based on the response of the candidate, the so-called "structured oral" is an essential part of creating that plateau of equality that is so important to all candidates. Therefore, a good oral board will have a game plan for asking the questions and a time system so that all the issues will be addressed and each candidate will be given a fair opportunity to present his knowledge, skills, and abilities. As a candidate, if you understand what the board is trying to do, you will do a better job in responding to the situation.

As mentioned before, boards look for two things: physical behavior and verbal responses. The more an evaluator writes down what he sees and hears, the better he will be able to make a fair assessment of the candidate once the interview is over. Writing a shorthand version of behaviors and responses to questions is also a skill. Some evaluators simply cannot listen and write at the same time, but those with more highly developed skills in taking adequate notes while listening and observing the candidate do a better job of giving that candidate an appropriate ranking.

Board members in some departments take photographs of candidates with a Polaroid and post them on a board at the end of the day to refresh their memories of each person. Personally, I prefer to make a few short notes at the beginning of each interview about the candidate's dress and any interesting, distinguishing physical traits, such as a moustache, hairstyle, or type of tie or attire.

Observing behaviors means writing down things such as how well the person keeps his composure during the discussion. A person's posture and physical reactions, including the use of hand gestures and eye contact, can all be written down. Any repetitive physical gesture, such as constant tugging at a piece of jewelry, clearing of the throat, or twiddling of thumbs, can be written down, although these behaviors are not always negative; they usually express a form of anxiety and nervousness. The use of different types of gestures to indicate understanding or contemplation and other types of body language can indicate a person's type of personality and his reaction to the examination process.

Generally speaking, however, body language is not as important as the verbal responses. While we often use body language as a backdrop to form our initial impressions of an individual, we have to be extremely careful not to let physical presence detract from what an individual is saying in response to the questions. The primary thing we are looking for in response to an oral question is a comprehensive, well-organized, and appropriate verbal response. Also, when asking the questions, we must be careful not to give them particular twists by using humor or slang. A straightforward question allows each candidate to make the most of the opportunity. Providing variations on the question for individual candidates can create a dilemma for the evaluators at the end of a day in comparing one response against another.

Keeping shorthand notes of what the person says in response to the questions, and even describing the person's behavior as part of the response, is appropriate. It is not uncommon for people to hesitate between the time you ask the question and their response. This may indicate different things. If a person hesitates on every single question, it may be an indication that he is very cautious in his response; and if his answers are comprehensive and well thought out, his hesitation may be appropriate. On the other hand, long gaps of silence before responses that are superficial, may indicate that the person is stymied by the questions because he lacks depth in those areas.

Seasoned evaluators seldom base any evaluation on one single response to one single question. What you are looking for is the total person in the context for which he is being evaluated. It is not uncommon for a person to be slightly nervous at the beginning of an oral examination and for his confidence level to build as he responds to questions. On the other hand, candidates have been known to act extremely cocky and almost arrogant when they come into the room, but leave the room acting as if they had just been whipped.

The job of the evaluator is to observe the candidate from beginning to end. The more comprehensive an evaluator's notes are regarding an individual, especially if there is any change in behavior or any distinction in response to different categories of questions, the more likely it is that an evaluator will be able to give a well-considered ranking to the person.

Unless the personnel rules require otherwise, it is not a good idea to try to place the first one or two candidates of the day into a numerical scoring mechanism. While it is a good idea to keep notes on each person, if you start early in the morning by giving the first person extremely low or high scores, you can put yourself into a difficult set of circumstances as you observe the performance of the remaining candidates.

During one oral board, as the first two or three people of the day were interviewed, they looked really good. One of the evaluators gave all of them 90s. As the day wore on, however, we discovered that those first people were the lower candidates; the new individuals who came in were even better. This required making erasures on forms and amending scoring, which is a bad practice for several reasons. Frequently, personnel rules require that individual candidates be allowed to read the remarks and scores of the examination process. Erasures and strike-throughs always bring a certain amount of discredit on the evaluation process. You should not be too hasty to place a person into a category until you see how that person's responses measure against those of his peers. This admonition applies whether you are testing candidates from the same organization or whether it is an outside examination involving people from other organizations.

One concern faced by evaluators is how to write up the average candidate. Average is usually just that—middle of the road. If candidates have an average response and that is adequate, it deserves to be called that. In the competitive process, you are basically looking for two specific types of candidates: the high-level achievers and those who should not be considered for appointment. The distinction between a good evaluator and an above-average evaluator is often the evaluator's ability to discriminate in this area.

If a person is above average, the evaluator should be able to give specific reasons why the individual appears to be that way. Statements like "I think this person will make a fine captain" are not adequate. On the other end of the spectrum, failing a candidate with a statement such as "This person should not even be considered for appointment" is also inadequate. What is needed is specific justification for above- or below-average performance. If the individual is above average, has comprehensive responses to all questions, maintains good composure during the testing process, and has the ability to respond to follow-up questions as easily as he responds to the primary questions, then that is what you should say. If the individual seems extremely anxious, displays body language of indifference by slouching in the chair and failing to maintain eye contact with the evaluators, provides superficial responses to questions, and creates long periods of silence, then you should say that.

Do not underestimate the value of giving a comprehensive response to an above-average or below-average candidate. Even above-average candidates often have some kind of weakness on which they need to work. I have had many opportunities to discuss my evaluations with candidates in the aftermath of their oral boards, and I have found my written comments to be extremely useful in giving them counseling and guidance. This is a tenuous area that some evaluators shy away from entirely, but I am of the opinion that if you have the ability to do good evaluations, you should not be embarrassed by either recommending or failing someone. I consider the written commentary to be useful information for training and, in some cases, giving feedback to the organization about the deficiencies of their departmental training program in preparing candidates for the job.

At the end of a day spent as an evaluator on an oral board, you will have examined many individuals. Depending upon the scheduling, the different responses to the questions begin to take on a fuzziness. That is why your written responses at the time of each candidate's involvement are so critical. It is a good idea at the end of the day to lay out the forms in the sequence in which the candidates were interviewed. This will refresh your memory about the sequence of events of the day. Then it is a good idea to lay out the forms in terms of the ranking of candidates. This will help break through the monotony of the interviewing process as you try to recall the best and worst candidates. I find it extremely useful to identify both ends of the spectrum first and then work my way back toward the middle of the range. It is appropriate to add additional comments at this time, and you should also make sure that you have provided adequate justification for your ranking decisions.

Some organizations hold a debriefing for evaluators with the personnel officer. While this sometimes focuses on candidates' qualifications, it is also appropriate to discuss the organizational context. I once sat on an oral board where we failed all the candidates. During the initial briefing by the personnel director and the city manager, we had been given specific direction regarding the type of candidate they were seeking. None of the individuals complied with those needs. The debriefing that afternoon consisted of talking about the organizational expectations and the recruiting process.

If you are like most evaluators, you have had to take some kind of oral board examination in the past, and that should give you a certain amount of empathy for the person on the other side of the table. On the other hand, if you are an incumbent in a job to which others are aspiring, you have an obligation to make sure that the candidates meet the standards. We are not doing candidates a favor by giving them encouragement based on faulty premises. If someone is not up to the job, the person needs to be told that this is the case. The process itself should be the basis for the elimination.

Good evaluators have the ability to provide a description of a candidate that is a unique blend of specifics and generalities. Poor evaluators can only express dissatisfaction with a candidate for general reasons, for example, by stating "I just didn't feel right about the guy," or making some other trivial statement.

THE ASSESSMENT LABORATORY PROCESS

The type of examination process known as the assessment laboratory is being used more and more often to screen candidates. This is both good news and bad news for candidates. On the one hand, the process is more job-related than other processes. On the other hand, some people who are experts at taking the test do not perform that well upon appointment. Does the assessment laboratory process locate the best candidate? The answer is probably sometimes yes, sometimes no.

What does that mean to you? If you have been adequately prepared for a job, you are going to do well in the assessment laboratory process without too much coaching. If you are not prepared, someone else who can act the part may be chosen over you. In the first case, the process has worked, but in the second case, the department that is looking for a competent person has lost. Many people take classes on how to pass an assessment laboratory but overlook one key point: the best way to do well in the process is to be well prepared to fulfill the job.

And that is my suggestion here: if the process of selection includes an assessment laboratory, the first thing you must do is "become the chief."

Approach the entire process as an opportunity for you to demonstrate what you would actually do if you were confronted with the tasks after appointment. This is not a game but, rather, a test of credibility. Approach each exercise as though it were reality. Do not worry about your score. Worry only about whether you are doing the task the way you would do it if it were "for keeps."

Now, you need to realize that this tactic can be both good and bad, since the assessors will, in fact, be scoring you. If your choice of techniques is good in their perspective, you will get a good score. If your techniques are bad in their perspective, you will not get a good score. But, in both cases, the assessors will be scoring *you*, and that is the most important part. Assessing candidates is not a science, but acting out a part is not science either. Assessment laboratoriess, if they are good, will pick the best candidates if all the candidates are telling the truth by the way in which they are performing.

The absolute worst thing that you can do is to "perform" on a test by exhibiting behaviors that are not representative of the way you think, feel, and behave in the real world. I once had a personal experience that really brought this home to me. When I failed an assessment laboratory process, I asked one of the assessors for some feedback. His response made me feel very positive about my personal choices. He told me that when I was participating in several of the exercises, my ideas and proposals were looked upon by the assessors as being "too radical." Yet, they were my ideas and proposals, and they were based on things I truly believed in. I did not get that job, but I am glad that I did not. Later, I was told by a reviewer in another test that one of my answers was almost identical to the answer given by the personnel director. In the former scenario, I did not get the job for the right reason. In the latter case, I had the potential of being able to assume a positive relationship fairly quickly, and I was offered the job.

In assessment laboratories, be yourself. Say what you think. Act like you think you would act in real life. Advocate positions you believe in. Write your answers as if they were answers to real problems. Present your ideas as you would in a real staff meeting or council session. If you do this, I can almost guarantee two things: if you are what they want, you will be given the chance to prove it, but if you are not what they want, you do not want to be there anyway. The worst of all worlds is to be one thing for a selection test and another thing for the real test—doing the job.

The testing process is a crucial point for you as an individual, but it is also a test of the evaluators and the department at the same time. If you, as an individual candidate, put your best foot forward, you should be among the finalists. Just remember, the race does not always go to the swiftest, but to the best prepared and the most persistent.

SELECTED READER ACTIVITIES

1. Develop a résumé file that contains actual records and evidence of your past experiences and performances.
2. Develop a generic résumé that tracks your career in reverse.
3. Obtain a copy of a job flyer for a fire chief's job for which you are interested in applying. Compare the attributes being sought in candidates according to the flyer with those in your generic résumé. Prepare a customized résumé for the specific job flyer.
4. Actively seek to serve as an evaluator both on oral boards and in assessment laboratories for your agency or others.
5. Mentor a fire officer who is one rank level behind you in the preparation process. This might include reviewing his résumé and conducting mock oral board and assessment laboratory processes.

The

Transition

Process

MOVING AND MOVING AGAIN

CHAPTER 8

Transition, Transplants, and Training:

Moving into the Chief's Seat

The object of this chapter is to talk about what happens when a person promotes from within an organization and what happens when a person comes in from outside an organization. Emphasis is upon the assets and liabilities of both situations.

When in chariot fighting more than chariots are captured, reward those who take the first. Replace the enemy's flag and banners with your own, mix the captured chariots with yours, and move them.

131

> *Treat the captives well, and care for them.*
> *This is called "winning a battle and becoming stronger."*
> —*Sun Yat Tzu,*[1] The Art of War

RIPPING OUT THE ROOTBALL: MAKING TRANSPLANTS WITHOUT KILLING THE PLANT

You have received a phone call or a letter offering you a job. Now there are things that you need to think about before you call and accept the job.

If you have ever had a garden and had to transplant something from one container to another or from a seedbed to a permanent position, you know that transplanting requires a specific technique. It requires some attention to detail, since one possible outcome of transplanting something is mortality of that which is being transplanted. When a person moves up within an organization or from one organization to another, that is, from a position of comfort in an organization to a competitive position involving upward mobility, this same process applies.

If you become a chief officer by coming in from the outside, you will not have a taproot. If you move up from within the organization, the taproot may or may not remain healthy. In either case, you cannot count on the nurturing elements in the organization to make sure that your administration will grow healthy and strong. Therefore, you must pay close attention to the transplanting process to ensure future success. If you grew up in the organization, you are no longer going to be able to rely upon past experiences for nurturing. If you have come in from the outside, beware that there may not be much in the way of support.

In the wonderful world of gardening, three things make transplants work very well: making sure that the plant is transplanted at the appropriate time, making sure the soil is adequately prepared to receive it, and staking the plant so that it has a chance to grow straight and strong. The process of making the transition from a lower rank to the position of fire chief involves similar steps: a conscious decision to make the move, preparation of your family, and maintenance of your support network.

Let us talk first about the conscious decision to move into the new position. Some people are so enamored with the badge that they will move up for any reason. In a previous chapter, you were asked to determine your reasons for wanting the chief's job. Now they will be tested. Upward mobility

---

[1] Sun Yat Tzu, *The Art of War,* trans. Thomas Cleary (Boston: Shambhala Publishers, 1988).

as a goal in itself is not healthy. Sometimes a person can make this move looking for a symbol of achievement instead of an opportunity to achieve something of substance. One of the first things to take into consideration when you are getting ready to make an upward move is what your growth potential will be in your new position.

Both evaluators and candidates have made comments during oral boards about how difficult it is for a person to go from a very small organization to a medium- or large-sized organization, and vice versa. I do not agree with the idea that a person with experience in an organization of one size is going to be restricted in performing in an organization that is a different size. Individuals who take a comprehensive approach to their role in smaller organizations and who realize that upward mobility to a larger organization merely gives them the opportunity to expand are often more successful than people who make lateral transfers. Experienced individuals from large organizations can bring a depth of knowledge that can really benefit a small organization. But, most importantly, neither type of individual will do a good job unless well prepared for the transition. The conscious decision to be uprooted and move to a new organization or to be promoted within an organization should be based on an introspective analysis of how much growth you anticipate by taking the new job.

Internal candidates have to worry about whether they will be an "heir apparent" or a "new broom," both of which have certain implications. An heir apparent often assumes the job and merely tries to sustain the administration of his predecessor. A new broom is often expected to clean house. The problem is that when an heir apparent starts to make a lot of changes that were unanticipated, forces of resistance arise; and when a new broom goes for weeks or even months without making any changes, the appointing authority starts to wonder what is happening. The key is to determine the expectations the appointing authority holds for the internal candidate. A new chief from the inside had better know what is expected from above and what is expected by the department itself.

A person who is contemplating becoming the chief of a different department needs to ask some key questions. If the job looks like your current position, then where is the growth? If the organization has, and wants to maintain, basically the same kinds of programs, projects, and services, what is the challenge? A person who is considering a move to the top job in his own department also needs to ask some questions. What is the culture of the organization? Can you actually make changes in a system that you have helped to create? Will you be able to assemble your own team or will you just inherit the one that already exists? What are the opportunities to enact your own agenda?

The point of this kind of introspection is not to talk yourself out of making the transplant but, rather, to focus on the reasons why you are doing it. A decision to move up must be based on a realistic appraisal of the potential opportunities for your own professional development.

Our second element in the transplant process is preparation of your family for the change. Frankly, many individuals accept promotional opportunities for their own reasons and fail to realize the significant impact the change will have upon their spouses, children, extended family, and even their closest friends. You have to recognize that you may be giving up certain things in order to get something else. The strength of your relationship with the members of your family is extremely important at this time. You must have some empathy for their problems rather than demanding that they yield to your needs.

A classic example of the impact a promotion may have on a family occurred for a person I knew who wanted a specific career change very badly. It required that he make a major relocation, but his daughter was on the verge of graduating from high school. She was a top-level student and had gone to the school for a lengthy period of time. In this particular case, the solution was simple. Close family friends took the daughter into their household and served as a surrogate family while the father and the rest of the family moved on. This solution would not work for everybody, but the example illustrates the types of questions that we must consider. Once we decide to make a transition to a new organization, we have to openly discuss the effects of that decision with our families.

You need to develop a checklist that examines what your family has now versus what your family may have to deal with in the future. The classic example might be an examination of the educational opportunities: what kind of schools are available now, and what kind of schools will be available in your new job? Social connections such as church, fraternal organizations, and even professional organizations often can assist in the stabilization process because many of them have branches or chapters in different communities.

One thing that may be a surprise for many people is the idea that when you promote from within the organization, you will lose a lot of the same relationships that would be lost if you were to move on to another department. Do not ever forget that when you are the chief, you are different than you were when you were not the chief. I have seen many individuals devastated by the realization that they are no longer just one of the group when they accept the top job. I am not suggesting that you are going to walk away from them, but once you become chief, you will have to walk some places where the individuals in your department will not want to walk with you.

Some people can handle this transition without any repercussions, but they are in the minority. Most people experience a change in personal relationships when they accept the promotion. What is important to remember is that you should not be disappointed by this phenomenon; rather, you should work as hard as you can to not create circumstances that will accelerate the process.

I am sure that you have heard the old axiom about not burning your bridges as you cross them. Maintaining a strong working relationship with those who are in your previous positions is an extremely good idea, but never to the point of compromising your ability to carry out your job. If you are moving to a new oreganization, be aware that it is likely that those in the new organization will contact their counterparts in your old organization to ask what kind of person you are. If you left today, what would the answer to that question be? When one exits gracefully from one organization and moves on, one's reputation not only follows, but usually is embellished a bit. Part of the transition process is to make sure you leave without that reputation being damaged by last minute conflicts.

When moving into a new area of responsibility, be aware that professional associations are always anxious to welcome new members. It will help during the transition period if you establish yourself as quickly as possible by attending their meetings and becoming involved. They become part of your new infrastructure, and that may be important to your integration into the culture of the department. If you already have a few friends or professional acquaintances among such groups, now is the time to renew those relationships. Spend some time with those people, letting them show you the ropes.

Tearing up your roots and moving to an entirely new location is somewhat difficult, but it can be done with a minimum impact if it is a planned process rather than a series of accidental events. Spend time preparing yourself and your family, have a good idea of what you plan on accomplishing once you have to set down new roots, and build a support system around yourself. This is a different process than the one used in the competitive aspects of pursuing the job, but if you wish to enjoy the fruits of your future labors, your time on this process will be well spent. The roadway to success in the fire service is littered with the remains of individuals who did not take the time to work on this element of career development.

NEW KID ON THE BLOCK

If you were born and raised in one neighborhood, you were lucky. You probably had an opportunity to gain friends at an early age and go through many trials and tribulations of life with them. On the other hand, if you moved from one neighborhood to another or from one city to another, you probably experienced the trauma of having to make new friends and establish new relationships. Often the change was traumatic, although in retrospect, there are probably many things you would have done differently if you had the insight and maturity you have now at a young age.

Becoming the new fire chief is an experience similar to that of moving as a child, especially for those who move from one organization to another. These individuals have not had the opportunity to engage in the maturation process inside the fire department that fire chiefs promoted from within the ranks have had. The pathway to establishing new relationships has many obstacles and pitfalls.

One of the first things that a new chief officer has to recognize is that every organization has its own corporate culture. Regardless of your individual technical expertise, political savvy, or personality traits, you will have to deal with the corporate culture of the organization. It cannot be ignored. It cannot be changed overnight. It will not go away merely because of a change in top level administration.

THE ONE HUNDRED DAY WAR

So, you have accepted the job. You have considered the ramifications of the transplant, and you are ready for the first day on the new job. When you took the tests and survived the interview, you gave your best answers. Are all these answers going to work now that you have accepted the new position? Sometimes the real test begins only after you have accepted the position. The term *trial by fire* gains new significance once you have pinned on the fire chief's badge.

The real test is not a paper-and-pencil examination but, rather, a test of credibility. The organization is wondering whether you are the right person for the job, and you are wondering whether you will be accepted by the corporate structure. A lot of clichés come to mind during the first few weeks of a fire chief's tenure. For instance, "It is better to keep your mouth shut and be thought a fool than to open it and remove all doubt" has some merit. The first two to three months on the job are fraught with a number of booby traps.

A friend of mine, Chief Martel Thompson, who became the new chief of a department about the same time that I took over a new department, coined a term for this period. He called it the *one hundred day war.* He did not mean warfare in the sense of going into battle with the troops; rather, he was referring to the fact that this is the period when you make or break your image in the organization.

This one hundred day war is a cold war—a war of wits, negotiation, and diplomacy. It is a critical period in the development of new relationships as fire chief. While every new chief's experiences are unique, they do seem to follow a general pattern, whether the chief is brought in from outside the organization or is promoted from within it. The only difference is that the booby traps are better hidden from the outsider.

The first booby trap is encountered early. You can expect to be presented with a list of all the perceived inequities committed by the previous administration and a request that you resolve them. The more pent-up the perceptions of these inequities have been, the more rapidly you will be expected to provide "reasonable and expeditious" solutions. You may be tempted to provide the requested changes in order to ingratiate yourself with the grievants and earn a few pluses as a decision maker. But remember this admonition: Behind every complex problem, there is a simple answer, but that answer is probably wrong.

The next booby trap is the department confidante. When you are the new kid on the block, there is always someone in the organization who wants to serve as your confidante, an organizational version of the "pipeline." That person will provide you with inside information and expect you to reciprocate. In exchange for the local gossip, the individual will want to know about your plans for the organization. A word of warning: this person, who claims to be your best friend today, will not be there when the going gets rough. In addition, the individual is likely to tell others what you are planning long before you are ready to reveal your strategy. It is prudent to make enemies slowly, but to make friends even more slowly. You have to learn to trust people, but make sure they deserve your trust before you expose your thoughts too freely.

The third booby trap is one you may unwittingly set for yourself. As you look over the new organization, you will immediately recognize ways of doing things that are different from the way things were done in the department or organization to which you previously belonged. Beware of the temptation to engage in what I call "yustas," that is, telling your new staff how things "yusta be done" back in Kokomo or Timbuktu, or wherever you came from. Just remember that the way things are done anywhere is based on what works there; and pounding square pegs into round holes will wreck both the pegs and the holes. What worked where you came from *may* work in the new place, but it must be properly evaluated in the new context.

As a young boy, I worked with my grandfather in setting up beehives. He showed me the technique for introducing a new queen bee into an existing hive. If a new queen is unceremoniously dumped on the bees in the hive, they will get angry and kill her. So we would put a queen bee in a wax shell and place the shell on the entrance to the hive. The bees would begin eating their way through the wax. This process took time, but as they removed the wax from the entrance, the workers became accustomed to the new queen's presence. By the time they were finally able to enter the chamber, they had accepted the new queen.

My reason for telling this story is to stress the importance of an adjustment period. The reason the bees accepted the new order was because they had time to adjust. During the interval in which the workers were chewing through the wax shell, the queen undoubtedly was communicating with them. The shell made the encounter less threatening. If you are fortunate enough to be named a new chief, you would be well advised to consider a similar process of adjustment.

The ground rules for an adjustment process are relatively simple. They are as follows:

1. Issue a directive immediately upon appointment to assure all members of the department that the practices and procedures already in place will continue until you advise differently. I call this the continuity-of-administration or change-of-command directive.

2. Immediately start a notebook of observations, based on what you see, hear, and read, that are pertinent to your understanding of how the organization currently functions. This is a form of "honeymoon" diary.

3. Conduct interviews of the organizational staff from the top down. Be especially careful to interview those whom you perceive to be troublemakers in the organization as well as those who you think will be supportive of your administration. Nobody should be able to tell which category you would apply to which individual. Listen twice as much as you talk. Look for trends and patterns. In really large organizations, it will be almost impossible to include everyone down to the firefighter level, you can talk with at least some of them.

4. Remember the old adage, "If it ain't broke, don't fix it." Do not change anything unless you absolutely have to change it. More often than not, the things that can be changed quickly come back later in the form of more complex problems. Make sure you thoroughly analyze and justify every change you make.

5. Introduce concepts to your staff as ideas and practices in which you truly believe, rather than as policies and procedures that you have brought along from previous employment. Help your staff access your network and the information sources that have led you to embrace and use these concepts.

There is another aspect to these recommendations. Mao Tse Tung said, "If you must do evil, do it all at once." A new fire chief might interpret this to mean that if he must make dramatic changes, he should make them all at once. Sometimes a person must start his administration by cleaning house. If you have done your homework before accepting the job, you will know whether this is a possibility. If so, chances are that the results will be more warlike than you have anticipated.

As a new chief, you are supposed to be building something, not tearing it down. The process suggested in this chapter will not ensure your success in your regime as chief, but it will give you a reasonable amount of time in which to prepare a game plan. The first one hundred days will be a difficult time, regardless of the situation. There will be some who will be disappointed because you are not rectifying old ills or meeting their expectations of you as a proponent of change. The payoff begins, however, when you are able to bring about changes that will last and be meaningful to the organization.

Removing the *causes* of grievances takes more time than merely adjudicating the grievances. The one hundred day war is more like guerrilla action than trench warfare; it is a period for gathering combat intelligence, not body counts. If you implement the suggestions provided here during those crucial first three months, you will increase your chances of establishing a solid footing in the organization. After all, you have nothing to lose but the war.

There are some things that fire chiefs should avoid while integrating themselves into the new corporate culture. There are some obvious corollaries to these *don'ts* but it is useful to look at these rules in a negative fashion because they have tremendous negative implications if they are violated. Simply stated, the ground rules of integration into a corporate culture are as follows:

A. Don't make assumptions.
B. Don't make promises.
C. Don't make threats.
D. Don't respond to rumors.
E. Don't speak about the culture until you know the taboos.

Assumptions can get you into deep trouble very quickly in a corporate culture. Assumptions that you know why things are the way they are or how they got to be that way are often superficial and inadequate to prepare you for establishing strong relationships in an organization. Many a chief officer has been derailed in his first couple of months in an organization by assuming that the fire department was incompetent prior to their appointment or that all the department wants is a "firm hand" to lead it out of the wilderness. In most cases, fire organizations have both strengths and weaknesses. The corporate culture that exists in most organizations is based upon the perceptions that grow out of those strengths and weaknesses.

It is all right to make observations about what you believe is going on in the organization, but you can get yourself into considerable difficulty if you base decisions on your own assessment of the situation. It is often a good idea, especially if the organization has long-tenured employees, to spend a considerable amount of time obtaining a historical perspective on the organization. This does not mean that you cultivate yourself to accept things the way they are; to the contrary, you may still have strong value judgments regarding the rightness or wrongness of any specific activity in the organization. But there is often an evolutionary element to the situations that is synergistic.

What this means is that, in an attempt to solve one problem, you may open other issues and controversies in the organization. The original problem may be irrelevant but may deplete the energy of the organization nonetheless. This is especially true if a new chief attempts to introduce a large number of changes in a short period of time. The assumption that a fire department's organizational structure is not functioning correctly or that some program has somehow been inadequate may lead the new chief to propose changes that have historical precedent in the organization.

For example, one recently appointed fire chief attempted to evaluate his new department's Incident Command System (ICS). The lack of an adequate ICS in the organization had drawn his immediate attention. His assumption was that the department was ill-informed and not prepared to utilize ICS. Without conferring with the staff, this individual issued a new operating procedure to the organization that was merely a photocopy of the operational procedure from his previous fire department. What this chief did not know, however, was that the department contained several highly qualified individuals who had served as instructors for ICS and who had been lobbying the previous administration to adopt ICS for years. In their respective shifts, these officers had been utilizing a mini-version of ICS when handling their own emergencies, since the former chief had never listened to the radio traffic or become involved in fire ground operations.

One might assume that the chief's issuance of a standard operating procedure that embraced ICS was greeted with a great deal of enthusiasm in the organization. To the contrary, the individuals in the department who had expertise in ICS had learned a specific type of command system, based on their department's needs, that was a relatively simplified system. The agency was smaller than the one from which the chief had transferred, and it had fewer demands for the sophisticated elements of the ICS. Instead of having a positive effect, the new standard procedure had a negative effect.

Resistance was rampant, even among the individuals who supported the concept of ICS, because it was force-fed into the organization rather than being introduced into the organization.

Making promises is the second area in which a new chief executive can be derailed. When one enters a new organization, the individuals in that organization often take this opportunity to purge themselves of frustrations, anxieties, and ill feelings toward the previous administration and their peers or subordinates. The new chief officer has to be extremely careful to not make promises during the initial phases of transition. There are several reasons why this is so. First and foremost, you may promise something that you cannot deliver. Secondly, you may promise to deliver something that runs contradictory to the interest of other people in the organization and, in the process, polarize the department. Even the most meaningless or trivial promises can have tremendous effects if they remain unfulfilled in the minds of the corporate mentality.

An example of this was a chief officer who, during the selection process, visited the fire stations and told the firefighters that if he were selected as fire chief, he would resolve many of their personnel issues. In the attempt to gain favor with the labor group, the individual failed to realize that the group's memory was far longer than the coffee break. Shortly after his appointment, he was approached by the president of the Firefighter Association who asked the chief when he would deliver the promised benefits. In an attempt to forestall the inevitable, the chief deferred and asked for an increment of time so that he could "do his homework." When that increment of time elapsed, a return visit by the president of the association was a disaster. The chief admitted that he had not succeeded in acquiring the benefits and tried to blame the failure on his superior. Who do you think got the blame at the next meeting of the Firefighter Association?

Promises are verbal contracts that build expectations in individuals and in organizations. When entering a new corporate culture, one has to be concerned about the conditions in that organization, but making an open statement about the immediate resolution of any specific circumstance is almost suicidal.

The third issue has to do with making threats. One might think that a threat is exactly the opposite of a promise. A promise is a contract to deliver. A threat is an action to damage. This is especially dangerous for the new chief officer in an organization that has any degree of polarity present when the chief arrives. "Choosing sides" from the outset of an appointment can drive a wedge in the organization that may be irretrievable. There is a fine line between choosing a direction for an organization and threatening the safety and security of the individuals within it.

Any student of Maslow's hierarchy of needs realizes that most organizations tend to function at the social level as more individuals within the organization gain tenure. A new chief officer who makes statements regarding the elimination of specific practices, changes in the direction of major programs, or restructuring of the social fabric of an organization threatens the very livelihood and "pecking order" of the corporate culture. Once these threats have been introduced into an organization, the individuals usually drop from the social level to the safety and security level in Maslow's hierarchy. Instead of making the resolution of issues easier, threats actually coalesce the individuals in the organization into pockets of resistance.

Rumor mongering is usually rampant when a new administrator takes over, whether from an internal or external promotion. The previous warnings against making assumptions, promises, and threats lead almost naturally to the recommendation to avoid reacting to rumors. In many cases, rumors are nothing more than the manifestation of certain individuals' assumptions regarding you as a person. Attempting to dispel the rumors, especially on an individual basis, creates the perception that you are protesting too much. Instead of removing the rumors, attempts to dispel them often lend credibility to them because of the amount of time and attention that they have attracted on your part.

I have worked in the command structure of five separate fire agencies over the years. In providing administrative direction to one department, I was involved in the development of certain programs and activities that fit the needs of that organization but were not necessarily appropriate for my next assignment. Upon assuming my new command, I was greeted with numerous rumors about changes that individuals predicted I would be introducing into the new department. Most of these were carryover actions from my previous department.

Obviously, most of us carry with us the wisdom gained from our experiences in our previous jobs, and in many cases, we would like to bring along practices, policies, and procedures that have worked well in the past. Protesting too loudly that you are *not* going to do something diminishes your ability to use whatever organizational transfer may be appropriate. If you attempt to dispel rumors by saying that you will not do something and then shortly thereafter introduce some practice that has been transferred from your previous organization, you will only be adding fuel to the fire: "See, I told you things were going to be that way!" So the best technique is to make sure you have completed your assessment of the department before advocating any change.

The last item in the list of "don'ts" centers on what could be referred to as the political faux pas. When you enter an organization, you have no idea who has a relationship with whom. The stranger walking down the street may well be the sister or brother of one of your superiors or subordinates. Therefore, you are well advised to not make any statements openly about the attitude, behavior, activities, or personality of anyone until you determine the backgrounds of the individuals in the new organization.

Probably the classic example of this situation was a chief officer who had decided to attend several council meetings prior to accepting an appointment as a fire chief. During one council meeting, a particular council member made several statements that were offensive to the person being considered for appointment. The prospective chief was sitting in the back row of the chamber, with only a few people nearby. Turning to the individual sitting next to him, who was another fire official, he made several critical remarks about the council member's comments. He noticed a woman sitting in front of the two of them, but she did not register any reaction to either the council member's or the chief's comments. Several weeks later, after the chief had received his appointment, he was invited to participate in a Chamber of Commerce mixer held in his honor. Imagine his chagrin when he discovered that the lady who had been sitting in front of him during the council meeting was the wife of the council member in question. Stories of this type, which are legion, illustrate how you can become entangled in a dialogue that is critical of individuals at the outset and end up burning some bridges before you ever get the chance to cross them.

All of the rules covered in this chapter highlight some of the potential pitfalls in each new chief officer's transition. The degree to which you avoid these pitfalls often depends upon your awareness of their existence. The process of becoming acclimated to an organization is similar to that of getting into a bathtub. It is far best to put your toe into the water and test the temperature before leaping bodily into what may be "hot water."

The process of acclimatization works. People have found ways to survive the frozen depths of the Antarctic and the scorched deserts of the Sahara. Fire chiefs can learn how to survive in organizations that are reluctant to accept them as well as in organizations that receive them with open arms.

SELECTED READER ACTIVITIES

1. Prepare a list of items that may be of concern to your family if you were to assume command of a department other than the one to which you belong today. The list may include, but need not be limited to, things such as schooling, doctors, personal friendships, and housing.

2. If you are a candidate for promotion within a department, prepare a list of items that could change as a result of the promotion. The list may include, but need not be limited to, things such as lifestyle, relationships, personal finances, and level of stress.

3. Prepare a hypothetical questionnaire that could be used to interview personnel after you have been appointed. What are the important things you would want to ascertain from your staff?

4. Conduct an interview of a fire chief who has recently completed the first year on the job. Ask the chief to provide you with an overview of the things the chief experienced, both things that the chief was prepared to deal with and things that took the chief by surprise.

5. Based upon your personal reasons for seeking the promotion, prepare a list of items that you believe would be important to review in your first year as a fire chief.

PREPARING FOR CRITICISM

CHAPTER 9

Suddenly a Shot Rang Out:

Preparing for Criticism from Others

The objective of this chapter is to describe the stresses and strains after the so-called honeymoon period is over. The chapter explains various strategies that fire chiefs can use to protect themselves from criticism as new chiefs.

Lessons from Parris Island
• *Tell the Truth.*
• *Do your best no matter how trivial the task.*
• *Choose the difficult right over the easy wrong.*
• *Look out for the group before you look out for yourself.*
• *Don't whine or make excuses.*
• *Judge others by their actions not their race.*

—Anonymous

CASTLE CONSPIRACIES AND OTHER COVERT CRIMES

Let us admit something right away. When you decided to become a fire chief, you made yourself a target. By adding that last trumpet to the badge, you stuck your neck out a country mile. You are going to acquire a lot of scars by being responsible for the department. Some of them will be on your forehead from butting it against brick walls. Some of them will be on your back. Some will be self-inflicted wounds, and others will result from attacks on you both personally and professionally.

Criticism is a fact of life for the fire chief. If you are a progressive fire chief, some people will do everything they can to slow you down. If you are a traditional chief, some people will try to speed you up. It is practically impossible to find a case of homeostasis in the fire service. Even if everything is going smoothly within your organization, there will be people outside it who will want you to do something different. Sooner or later, you are going to face a crisis.

A fire chief once asked me a rhetorical question, "What happens if you do all the right things and someone is working behind the scenes to destroy everything you are doing?" That is a really good question. We are educated in management and trained in fire fighting to do all the right things, but what happens if someone engages in a conspiracy to create a personal crisis for us in order to either destroy our credibility or blunt the momentum of our programs? I call these situations castle conspiracies. Where is it spelled out in an operations manual how the fire chief should deal with this type of problem?

In its most simplistic form, a castle conspiracy can be nothing more than a mild irritation. At the other extreme, fire chiefs have lost their jobs, seen their careers destroyed, watched their personal assets be bled away by attorneys, and faced a host of other unpleasant situations by failing to deal with these conspiracies. The question is, how do we deal with personal attacks?

Are there tactics and strategies that can be used by the fire chief to protect himself? The answer is yes.

The fire service does not have a monopoly on crises. At the political level, both the national and state level, crises in management have been studied extensively. At the local level, city managers and police chiefs have been dealing with crises in management for a much longer period of time than the fire service.

A book by Stephen Fink, *Crisis Management: Planning for the Inevitable,*[1] is a good reference for this type of consideration. Fink, who was a member of the staff appointed by the governor of Pennsylvania to the special Three Mile Island Crisis Management Team, developed an excellent approach to crisis management that matches the needs of fire chiefs very well. Fink describes how we can forecast crises, intervene to prevent crises from occurring, establish a management plan that includes a survey of the impact of a crisis, identify the factors involved, determine how to isolate the crisis, and, most important, learn how to survive "after the fall."

When I read Fink's book, I was impressed that most of the situations Fink uses to describe the behaviors of both successful and unsuccessful participants in crises involved emergency circumstances. The Three Mile Island situation, while it was primarily a hazardous materials problem, involved large numbers of emergency service personnel, including the fire service. The Bhopal incident was mentioned also. These types of crises are predictable and inevitable. They are a function of the technological society we have developed.

Other examples used in the book are not as obviously inevitable but they are going to occur. These include the types of "disasters" that affect chief executive officers of major industrial or commercial enterprises. The parallels in the fire service are the stresses and strains occurring in the implementation of affirmative action plans and code enforcement programs, and the labor relations type confrontations.

Fink points out that one of the key factors in dealing with crises effectively is knowing how to communicate under stress. Although a great deal of emphasis has been placed on communications under fire ground conditions, very few fire chiefs (or, for that matter, fire officers) have considered their communication skills when confronted with a battery of microphones and the all-seeing eye of the television camera. A person's demeanor in dealing with "bad press" or a hostile press may well be the one survival skill

---

[1] Stephen Fink, *Crisis Management: Planning for the Inevitable* (New York: Amacom, 1986).

absolutely essential to crisis management. Some of the best firefighters in the country have been brought to their knees by a television commentator who asked probing and penetrating questions that do not have simple answers. A fire chief's "bag of tricks" has to include communications skills under these circumstances.

One of the best books on this subject is *Testifying with Impact*[2] by Arch Lustberg. The book identifies some do's and don'ts that can be very important to an individual facing a hostile inquiry. The information in relation to testifying before press conferences and television cameras has real merit, emphasizing how to communicate in hostile circumstances. Much of the material discussed in the book can be related to situations in which you are confronting a group of antagonistic citizens or a group of angry employees.

Fink's and Lustberg's books both emphasize one important aspect in dealing with crises: having a plan of operation *before* a crisis occurs. Crossing your fingers for good luck and hoping you will finish your career without having to deal with a crisis is pure folly. It is difficult for us to accept the fact that sometimes our crises are self-induced. Moreover, we tend to deny that a certain situation is going to occur because most of us feel we can steer clear of such circumstances. But occasionally things will go wrong.

The chief officer mentioned earlier indicated to me that all the textbooks telling how to plan, organize, command, and control emergencies do not address the issue of how to deal with a disloyal subordinate with political leverage who systematically goes around you to persons higher in the organization. The situation is even worse when the individual who is your superior encourages and cultivates that relationship. This is a complex crisis, and I will resist the temptation to reduce it to a simple solution, such as "nail the so-and-so who is doing it!"

Instead, I suggest that if you face this situation, you immediately back away from the problem and attempt to depersonalize it. Do not get caught up in the crisis yourself. You should look at crises that occur in your professional life as being crises of position, not personality. So what if someone has the ear of your superior? So what if your superior encourages it? Your overall strategy should be to retain control of yourself so that you can control the outcome. Your basic tactic should be to outsmart the other person, not to change your personality or get into a confrontation that may have the inevitable result of termination or early retirement.

---

[2] Arch Lustberg, *Testifying with Impact* (Washington, D.C.: American Society of Association Executives, 1983).

Going back to Fink's methodology, he suggests that one of the best things to do with a crisis is to dissect it before wasting time and energy reacting to pressure. It is far better to analyze, bring structure to, and understand the anatomy of the crisis before you attempt to resolve it.

In recent times, there have been cases where fire chiefs have been attacked with lawsuits and the filing of civil and criminal actions. Newspapers and regional magazines have attacked the integrity and competency of fire chiefs; votes of "no confidence" have been made public regarding the behavior of fire chiefs. All of these are crises that, on the surface, can become highly personalized. A person's pride can be wounded and the pocketbook depleted.

The tendency in these situations is to fight back and fight hard. The worst thing that can happen to a new fire chief is to allow the situation to occur during the transition period. Usually the outcome of these kinds of pitched battles is casualties on both sides. The classic management model of win-win/win-lose/lose-lose can be applied. A second rule of thumb is summed up with "facts are whispers and lies are shouts." It is often extremely difficult for a person to refute statements, allegations, accusations, and personal attacks, but shouting, or its literal equivalent, is useless.

The tactic here is extremely simple: do your homework. Documentation is everything. When the opposition shouts, it is time for you to turn down the volume but stay on the right frequency. When any individual is accused, the tendency is for that person to shout back, "I didn't do it!" Denial is often considered prima facie evidence of guilt. Not unlike the small child caught with his hand in the cookie jar, a denial statement only heightens the third-party observers' suspicions.

Instead, when dealing with accusations, respond by asking for specifics. Before denying something, find out exactly what it is you are defending yourself against. When these kinds of events occur, you are going to be angry. You may become bitter, and you probably will be afraid. Friends may abandon you, and your own family may question you. So what is the tactic? Preserve at all costs your sense of self-worth. You are going to have to make an adjustment in relationships with people, but do not give up on yourself.

One rule of thumb related to controlling the outcome of the crisis: If you are defeated, retire with dignity, and if you win, remain gracious. The one thing that can never be taken from you as an individual is your personal dignity. A person can retire from the battlefield with honor, which often allows the person to return to the battlefield with reinforcements at a later time to win another battle. Conversely, a person who wins in a crisis must be very careful to not engage in "Saturday Night Massacres" in order to ferret out and punish everyone who did not line up to support that person's "winning" side. Crises have a way of destroying alliances and, at the same time, creating new coalitions that make friends of former enemies and allies of survivors.

The Boy Scout motto "Be prepared" applies to the fire chief. We are extremely well prepared for crises at the scene of emergencies. We develop an extensive body of knowledge on how to behave when we are trying to restrict a fire to the point of origin. So it follows that we should be good at handling personal crises also. We just have to prepare for them.

Ben Sirach, in the Book of Ecclesiastes in the Bible, stated, "From a tiny spot comes a great conflagration." Who knows that law any better than firefighters? Crisis or conflagration—the fire chief should be able to keep both under control.

ELEPHANT BULLETS OR MOUSE BULLETS? THE RIGHT AMMUNITION FOR THE JOB

A city manager once told me that everyone who is given a job of leadership has a theoretical revolver that contains six cylinders. Three of these cylinders are bored to carry elephant bullets, like those in a 357 magnum, and three of the cylinders are bored for mouse bullets, like the 22-caliber bullets in a rifle. The real trick, according to this person, is being able, throughout your career, to match the right bullet with the right problem. And when you are out of ammunition, you are in real trouble.

This city manager went on to explain that throughout their careers, most top-level managers face a series of crises. If they do not, they are not working up to their potential as change agents. Some crises are monumental and require a tremendous amount of energy on a person's part to cope with them. Other crises are relatively minor, but others perceive them as being very important, so they must be resolved.

The trick is to make sure that you never use a rhetorical bullet unless you have to use one. You must be careful to avoid firing a weapon at every problem. If you do have a problem that requires you pull the trigger, make sure that you match elephant bullets with elephant problems and mouse bullets with mouse problems. Any mismatch is liable to result in a catastrophe. If you use an elephant bullet on a mouse-sized problem, you will waste a resource. If you fire a mouse bullet at an elephant-sized problem, you will get trampled.

My interpretation of this city manager's story is that there are times when a leader has to judge whether a problem is worth making an issue over it. If the situation is important to the manager, then the manager must be ready to use power. If the problem is important to someone else but not critical enough to cripple the organization, then the leader may need to use persuasion. When you use power, do so in a way that makes it clear that you truly mean business. When you use persuasion to avert a crisis, it should to be based on the fact that the other person cares more about the problem than you do.

There are many scenarios in which these kinds of ideas are played out in the context of the fire service. As mentioned previously, when entertaining a discussion of decision making in the fire service, most people tend to think that the most critical decisions are made at the scene of emergencies, but I disagree with that. I think the decision to use mouse or elephants bullets is going to be made most frequently in areas of program advocacy, labor-management relationships, and jurisdictional and political disputes.

When it comes to actually using power or persuasion, there are no magic formulas to help you determine when one or the other is appropriate. The decision to use power or persuasion is your decision. You will probably need to use power when power is applied against you, and you will probably need to use persuasion when you do not wish to overreact. Perhaps it is appropriate to make the following considerations before choosing to either.

One has to be very careful to pick the things for which one is willing to stand up and fight. My wife calls these "To Die For" issues. Elephant bullets should be reserved for situations in which there is no room for compromise and in which the consequences of failure would be very serious to you or your organization. You cannot afford to bluff or posture in hopes that the other person will back down when the stakes are high. But when the hammer comes down during an elephant charge, you still have to aim accurately if you hope to survive.

You have to be equally careful to not let a mouse-sized problem become a liability by failing to deal with it. If the consequences of failing to deal with an issue will result in its growing into a larger issue, then you need to use all of your communications skills to resolve the issue as early as possible. When you use a mouse bullet, it should be more subtle—more like you did not want to do it, but it had to be done.

In reality, we are not in the business of pulling any triggers on real weapons. These are just analogies, but they have real counterparts in the shape of problems and crises that demand action. Happy hunting! This next group will probably be the group that requires you to use the elephant bullets.

THE POWER ELITE

In every community, no matter how large or small, there is a relatively small group of people that makes most of the major policy decisions for the remainder of the population. This group can be labeled the "power elite." The nature of the fire chief's job involves making decisions regarding recommendations for adoption by the authority having jurisdiction. If a recommendation does not conflict with the power elite, there is no problem. If it does conflict, however, there can be a consequence.

Therefore, it is prudent for a chief officer to go through an analytical process to determine which individuals make up the power elite before exposing his programs to their critique. Individuals have derailed their entire careers by making the assumption that because they were the fire chief, they were automatically part of that power elite. While it is true that many fire chiefs have developed strong relationships in the communities they serve, more often than not they are not part of the power elite. Instead, they are people in position of authority. There are significant differences between the power elite and those with the power of authority.

The power elite consists of individuals who have a large amount of influence over the direction of public policy without being appointed or elected by, or for that matter even known, to the public at large. The uninitiated often think that the people who are in positions of appointed or delegated authority, that is, the mayor, city manager, police chief, and fire chief, wield the only power of the community. The power elite is much more subtle than that. Granted, sometimes authority figures are allowed to circulate with the individuals who are part of the power elite, but the two groups are not the same.

The power elite is made up of people who, if they want something, can almost always obtain it. Likewise, if they want something stopped, they can almost always prevent its occurrence. In both cases, they use their power within the community without being visible.

The use of power by this elite group is expressed in distinctly separate ways that are often mutually contradictory. For example, the power elite can be malevolent or benevolent. These individuals can be very direct or extremely indirect in their influence. More importantly for you individually, they can be either your enemy or your ally.

Which individuals become the power elite? There is no good physical description of a person who falls into this category. These individuals can be either male or female. They can be rich or, in some cases, very middle-class. The key factor to look for in identifying the power elite is not the symbols of power, but rather the tools to wield power. In general, the power elite consists of people who wish to influence the decision-making process in a specific direction so that their own values will be sustained through public policy.

Some people might simply call this politics. The use of power, however, goes a little further. In some communities, the small group of people who fall into the category of the power elite may have totally different political values but still exercise their influence over decision-makiing processes with an amazingly consistent practice.

The first ground rule to remember about these people is that you may not know who they are until it is too late. This is why it is very important that a fire official engage in the process of analyzing the interactions of the power elite in the community as quickly as possible after appointment to the position of making key policy recommendations. Identifying the individuals will, if nothing else, help you to avoid offending them by accident. There are horror stories told about individuals who made snide remarks in restaurants or critical comments in a public place only to find out later that the people they insulted were some of the most influential people in the community.

The individuals who make up the power elite usually share several features that helped to make them powerful individuals in the community. First is tenure in the community. Senior members of the community usually wield a great deal of influence in the political processes. Another common factor is possession of material wealth or control over economic considerations such as land holdings or major businesses. Large land developers and individuals who control a significant part of the commercial or industrial base of the community may not have a badge of office, but they wield a great deal of power with their bankrolls.

Another form of influence that is common to these people is that they have been part of the political processes for lengthy periods of time and have gotten other people elected into positions of power. Neighborhood action groups and other community groups that have been responsible for the election or nonelection of local officials often contain members who are as powerful as the elected officials. If you want to see an energetic power elite group at work, go to a meeting of individuals who are advocating a recall election.

Another group that has a great deal of political influence consists of those people in the educational system and the religious network. Because of their ability to evaluate and estimate grass roots reaction to a variety of public policy issues, religious leaders and educational leaders are often sources of information to other people. Therefore, they can wield a great deal of influence in the political process.

What does the power elite do? If you recall that this group can be malevolent or benevolent and direct or subtle, you will realize that there is wide variety in the ways that people in influential positions operate. Basically, these individuals do everything they can to control the destiny of the community in the areas in which they have an interest.

The power elite is not above taking matters into its own hands when the group believes that people in positions of authority are not responding appropriately. The outcome of this behavior can be anything from the recall of a political adversary to constant harassment of department heads who fail to meet the personal expectations of the elite group. An individual in this group who is malevolent, transparent, and indirect may be difficult to ferret out under some circumstances. Such an individual is very likely to become your enemy. On the other hand, a person who is benevolent, highly visible, and very direct has a high degree of possibility of becoming co-opted as your ally.

Because these individuals often have a great deal of influence over the communication process in the community, they can distort or amplify information that is being provided through formal channels. The power elite commonly includes individuals who have access to the mass media communications system. Since they are capable of carrying their case to the public through this mass media, they can quickly become an asset or liability to any specific public policy direction.

How should we respond to the power elite? First, we need to recognize that the fact that the individuals are the power elite *does not* necessarily mean that they are right. In your capacity as a fire official, it is neither ethically nor morally correct to bow to the influence of such groups in a subservient manner. Instead, we must learn to identify these people as quickly as possible and develop a strategy for working with them or around them in order to achieve the goals of fire protection.

Identifying the member of the power elite is not a simple task. It requires both investigative work and the use of logic. One of the simplest techniques to utilize in developing a sense of the power elite is to merely keep track of which individuals appear at social gatherings in which there are people in authority. Often the people in attendance are not there to pay their respects but rather to keep themselves visible to the person in authority and to remind that person of their relationship. If certain individuals seem to appear under a wide variety of circumstances, there is a high degree of possibility that they are not just hangers-on but rather part of a network.

During such social events, be very observant and determine who talks to whom and under what circumstances. If you note that an individual is treated deferentially by a person with a lot of perceived public power, and if it seems that the individual is giving advice more than listening, there is a possibility that you are observing a member of the power elite at work. Of course, this technique has a particular implication for a fire official. You must be out in the community and engaged in these kinds of social contacts if you are to avoid becoming victimized by the social fabric of the community.

Another technique is so simple that it is often overlooked. You can simply ask, who are the power people? When you are in the process of becoming familiarized with an organization, for example, when you first assume command of a responsible position, it is reasonable to ask people whom they consider to be the most influential people in town. I have conducted a series of such interviews in making the transition from one department to another over the years, and I have found that some people are very aware of which individuals have true influence in the community while others do not have a clue. If a person's name is mentioned five or six times during different interviews, that person is someone you should investigate. It is time to do your homework on that person and determine their base and mode of operation.

In one community where I worked, I discovered that one person owned a great deal of property in the town. Although he had never been elected to a public position, it was clear that he had a great deal of influence over the individuals who sat on the council. As I began to explore his background and his means of financial support, I discovered that he was a major contributor to the political campaigns of these candidates. A little additional homework revealed that he had been a member of many social organizations in the community containing the visible supporters of his candidates. He remained invisible to most.

Acquiring knowledge about the members of the power elite does not mean that you should develop a game plan for dealing with each one individually, but if you do not know who these individuals are, there is a high degree of possibility that you can be ambushed. The process of investigative research to determine the power base in your community will give you a snapshot of the people with whom you will have to deal, and often you can avoid conflict.

While conducting an investigation of these people, determine their values and what specific "freaky buttons" cause them to react negatively. A person's value system drives practically everything he does. The members of the power elite test every decision against their value systems. The term *freaky button,* while not scientific, is a description of a very real phenomenon in the world of politics. Some people have a limited tolerance for specific behaviors; if you step on their "buttons," no matter how logical you happen to be, you will get a negative reaction.

If there is one thing that will cause the power elite to react negatively, it is surprise. Fire officials need to determine as quickly as possible how to communicate with these people about direction and activities so that they will not be surprised. But, as noted earlier, there is nothing wrong with suggesting something with which the power elite will disagree. Even though these people exert a tremendous amount of influence over public policy, for the most part, they do not have the ultimate ability to prevent something from happening if it is actually an idea whose time has come. You can either co-opt them and make them your allies, or you can neutralize them and remove their influence. You can almost always be assured, however, that if you anger them, it will make your job more difficult.

One of the worst things you can do is to try to expose their self-serving interests. Any time you cross swords with members of the power elite and attempt to discredit them by raising their level of visibility, they are going to engage in guerrilla warfare against you. After all, they do not have to reveal anything. There are no public disclosure and conflict of interest laws for people who are not elected. The more you try to drag them into the limelight, the more likely it is that they will retreat into the shadowy edges and move around to attack you from another position. Unfortunately, you have to be in the middle of the limelight to drag them there, and you can be blinded by it at the same time.

What do you do if you get into trouble with the power elite? As we all know, an ounce of prevention is worth a pound of cure; therefore, if you do your homework by determining who these people are in advance and communicating with them, often you will be able to stay out of trouble with them. Unfortunately, however, if we follow the dictates of the fire protection profession, we will find ourselves at odds with one or more members of this group at some point in our career. When that happens, you will need to have a strategy in place for your own defense. Failure to do so has resulted in termination, early retirement, or forced relocation for many fire chiefs.

If you do run afoul of these people, focus your entire energy on dealing with the issue that raised their ire rather than on attacking them. If there is any time in your career when you most need to be principled, open, and direct about what you are trying to achieve, it is when you have individuals attempting to sabotage you through the influence of others. If you have communicated the substance and content of what you are trying to achieve to a broad spectrum of the community, you will stand a far better chance than if you have spent all your time trying to influence the smaller power elite. Issues that ultimately go to the public domain are often out of control of the power elite. The members of the elite do not control the electorate; they only control the communications process in the community. Therefore, one of the best strategies is to make sure your communications network is as broad as possible and to share as much information as you can about your course of action and the rationale behind it.

Second, once you realize that you have run afoul of the power elite, document every single thing that happens to you. If you are attacked in the press, maintain a clipping file of the articles. If you are attacked through correspondence, keep a personal copy of every letter. If you are attacked verbally via telephone and personal contact, start a log book and keep track of everything that is said—who said it and when they said it.

Although these suggestions may take on connotations of being paranoid, they have a logical purpose behind them. Quite frequently, malevolent and invisible power people try to intimidate through the use of direct force. They are often convinced that if you will not do things the way they want you to do them, they must remove you so they can maintain their power base. If the situation gets to that kind of hardball, you will be better prepared to defend yourself when the ultimate blow comes if you have documented everything that has happened.

Fire chiefs do not wear white hats anymore. It is not uncommon in many communities for the fire chief to be looked upon as a potential threat to the freedom and independence of many in the business community. Therefore, the strategy of dealing with the power elite is to make sure that you have a strong defense in the event that the members of that group try to make you a combat casualty by your removal. In several cases, individual documentation of the sort recommended here was instrumental in getting judicial pressure to restore individuals to their employment and then provide them with an opportunity to move on to other jobs without the liability of a black mark against their names. I have also seen situations where the development of such a strong defense mechanism gave the power elite second thoughts about attempting to remove an individual.

Lastly, do not ever try to bluff these people. The people accustomed to operating in this arena are not easily intimidated. The worst thing you can do is attempt to threaten them with some force of action that you have no power to implement. The fire chief who finds himself between warring factions can be certain that he will be the one who gets most seriously wounded.

People who fall into the category of the power elite seldom wear name tags identifying them as such. They often are so subtle in their influence on the community that they must be ferreted out. Yet, every community has them. Your survival as a fire official will probably not depend too much on understanding the power elite unless you choose a course of action that will be contrary to that group's value system. If you identify these individuals in the early part of your tenure as chief fire officer and maintain a channel of communication to ensure that you are monitoring their behavior and perceptions of your role, there is a high degree of possibility that you will be able to avoid conflict with them. Failure to identify them and to examine the implications of your actions in the context of their desires can spell disaster.

WHEN IN DOUBT, DON'T!

Some days are going to be rougher than other days. Any day might be one of them. Maybe it will be next week, or next year; but such days will occur. They are the days when you will wish you could have stayed in bed.

A day like this might begin with a telephone call at 0430 hours from the local police department. They have arrested one of your firefighters for possession of marijuana. He was stopped for driving under the influence of alcohol, and a routine search of his vehicle revealed a sufficient amount of the drug to warrant his arrest. You might have been at a party at his home just last week.

Shortly after arrival at your office at 0800 hours, you are greeted by a phone call from the fire chief of a neighboring town. He wants you to hire his nephew who is sixth on the list. The morning mail has a letter containing two free tickets to a wine-tasting party hosted by a local contractor. After a department head meeting, you are approached by a member of the city staff who asks if you would look the other way on a small fire safety violation at his brother's restaurant.

As you are preparing to leave for lunch, a local apparatus manufacturer drops in and insists on taking you out. He invites you to be his guest at a country club next week for a golf tournament. Finally, as you are leaving to go home, thinking that it is all over for the day, you observe an off-duty fire captain driving down the street with his arm around the wife of an on-duty captain.

Granted, most of us have very few days that are quite that hectic, but most of us would much rather be fighting fire than dealing with problems of this nature. These are scenarios that require reaction without firm guidelines. Each scenario represents the opportunity for a person to break the law, bend the rules, or take a stand. These types of problems are the most difficult problems a chief officer has to wrestle with, because each issue involves a set of ethics.

The Greeks gave us two concepts: democracy, which is a form of government, and ethics, which is a form of personal conduct. The root word for ethics is *ethikos,* which means a standard of behavior. Many of the personal crises fire chiefs face derive from their own ethical behavior. Someone or some group of individuals takes exception to the manner in which the fire chief conducts himself and decides that the fire chief's behavior is not up to their standards. The result is conflict.

Several years ago, while preparing a course for fire officers, I developed a game called "Code of Conduct." It was fashioned after a game on the commercial market called "Ethics." In this game, an individual reads a scenario that has several possible outcomes and then selects one of the alternatives to deal with the problem in the scenerio. The selection of the alternatives is based upon the personal values of the individual. Once the person has selected one of the options, the other players attack the decision and try to discredit it in every way possible. After a period of time, the players vote on whether they agree or disagree with the person's choice. There are no right and wrong answers, only points of view.

The old cliché "One man's ethics is another man's controversy" is illustrated in this game. For example, if an individual strongly believes in the concept of honesty, that individual's choices will be shaped by that belief in a very definite fashion. On the other hand, if a person bases his values on expediency, honesty may be a matter of situational opportunity.

Fire departments probably have such extensive rules and regulations because questions of ethics have often been interpreted in the context of the organization. In one fire department where I served, we jokingly referred to different rules and regulations in the book by the names of the individuals who were disciplined or chastised and caused the rules to be formulated. In other words, rules are a form of controlling ethical behavior.

Unfortunately, rules do not cover all of the types of situations in which a chief officer finds himself involved. Concepts that apply to a chief officer are different than those which apply to the lower ranks in the organization. Examples include conflict of interest, misuse of authority, and conduct unbecoming to the position. No fire department rule book is comprehensive enough to guide a person through these types of situations.

What is ethical behavior in the context of the fire service? Can a person perform within the boundaries of the rules and regulations and yet be subject to criticism? The answer is yes. Does it require an illegal act to breach the ethical considerations of an organization or a community? The answer is no. You can be in the right and still be considered wrong. Ethics are values. A person can be operating strictly within his own rules and regulations and yet violate some value that creates the perception of illegality, even when it does not exist.

I recall an incident in which a chief officer quickly established a set of very negative values. My station was involved in fighting a fire in a manufacturing plant that produced fishing rods. We did an outstanding job of controlling the fire. The following Monday morning, the owner of the facility brought a handful of fishing rods into the fire station, left them in the dispatch center, and told the dispatcher that they were for the "guys who fought the fire." We had a rule against accepting any form of gratuity that had value. When the fire chief saw the rods and asked why they were there, he was advised of the circumstances. The chief officer demanded that the rods be removed to his office, and the dispatcher was told not to discuss the matter with anyone in the organization. Of course, rumors of the incident circulated around the station.

Several months later, during a chance conversation, the rod manufacturer asked one of the firefighters how he had enjoyed the use of the fishing rod. The firefighter informed him that the rods had been taken by the chief to be returned to the manufacturer. The businessman looked puzzled, and both parties were reduced to embarrassed stares. The situation raised some serious questions and left the perception that some form of unethical behavior had occurred on the fire chief's part.

Although this situation was relatively blatant, other sets of circumstances that appear on the surface to be relatively innocent can have connotations that are just as serious. As chief officers, we have to be careful to protect the image of the position by preventing the development of a perception of impropriety. This is easier said than done, and we in the fire service are certainly not the only professionals plagued with this problem.

The *Harvard Business Review* published an article by Sol W. Gellerman entitled "Why Good Managers Make Bad Ethical Choices."[3] The advice in the article can be summarized in one statement, "When in doubt, don't!" The examples Gellerman used came from the corporate world but could just as easily have been drawn from fire department experience. Gellerman's article dealt with bad ethical choices made by executives of such major corporate giants as Manville Corporation, Continental Illinois Bank, and E.F. Hutton, among others. In each example, Gellerman pointed out how most ethical considerations can be rationalized by the decision maker to the point where not only are they not unethical, but, according to the decision maker, they are appropriate and essential. Gellerman identifies four rationalizations:

1. A belief that the activity is within reasonable, ethical, and legal limits—that it is not really illegal or immoral;
2. A belief that the activity is in the individual's or the corporation's best interest—that the individual or corporation would somehow be expected to undertake the activity;
3. A belief that the activity is "safe" because it will never be found out or publicized—the classic crime-and-punishment issue of discovery; and

---

[3] Sol W. Gellerman, "Why Good Managers Make Bad Ethical Choices," *Harvard Business Review* (July/August 1986). Boston, MA.

4. A belief that because the activity helps the company, the company will condone it and even protect the person who believes in it.

There are basically three strategies a chief officer should consider in dealing with the problem of ethical considerations. These three strategies are (1) adoption of a code of ethics, (2) documentation of interests, and (3) open discussion of personal activity.

Adoption of a code of ethics is more meaningful than the simple phrase implies. For example, when a person becomes a doctor, he swears to uphold the Hippocratic Oath. When you become a firefighter, you swore an oath of office. Codes of ethics in most organizations are not really stated as such. In most organizations, ethical standards are considered to be the rules and regulations. However, they are not the same. Rules tend to be series of acts and inhibitions, whereas a code of ethics is a statement of beliefs. Many organizations have a code of ethics that could be adopted by a local fire fighting agency with minor changes.

Documentation of interests addresses a different set of circumstances. In many states, a person who reaches the level of fire chief must file a document of public disclosure of personal financial interests. In some states, this conflict of interest statement clearly identifies those areas in which the individual might have opportunity to achieve personal through the decisions he effects. For example, a fire chief who owns a fire extinguisher company in the community would have a difficult time avoiding conflict of interest if he proposed an ordinance requiring annual inspections of extinguishers. Further, these kinds of documents often provide an opportunity for a person to make a matter of public record the advantages that accrue to the job. For example, it is perfectly legal in many cases for a chief officer or firefighters to receive a pass to the local movie theater, providing that the pass is adequately documented and a value is placed on it in a public disclosure statement. Whether or not one accepts the free pass or utilizes it after acceptance is a totally separate issue.

Public disclosure documents create a concern in some people's minds because they believe those disclosures are an invasion of their privacy. Other individuals who do not have the same level of responsibility can avoid disclosing their personal financial dealings. The advantage to public disclosure statements is that they do not have to be public in the sense that they are printed, published, and distributed to each and every person in the community. Instead, they are a matter of legal filing, and in many cases, they serve absolutely no purpose unless a person is accused of improper behavior.

Last, but certainly not least, is the subject of openness. If a chief officer is considering engaging in a particular type of behavior, such as attending a golf tournament with members of a fire equipment sales company, then he should mark it on the calendar and discuss it with staff members. Let people know where you are and whom you are with. If you are asked to participate in an activity that has some possible problems, discuss it with your superior or another department head. Nothing dispels rumors as rapidly as open dialogue. One of the best friends a fire chief can have, when it comes to avoiding unethical behavior, is the city attorney.

Members of your staff should be well aware of your involvements, and it pays to listen to their concerns. Loyalty on these types of issues is a two-way street. If members of the staff feel that they are being kept in the dark regarding your activities or behavior, they may feel it is appropriate to deprive you of information regarding their behavior. This does not mean that we have to engage in a dialogue each and every day about our personal involvements. What we are talking about is creating an environment in which honesty and stability are established.

It is interesting that the two concepts we inherited from the Greeks denote contradiction. Democracy is based on the premise that people should have the freedom to do whatever they desire as long as it is not in conflict with other people, and ethics says that the group can set some standards to limit that behavior. Our flexibility in choosing lifestyles and leadership methods means that we may sometimes choose a course of action that offends someone else.

If democracy is the engine that drives our system, then ethics is like the steering wheel. Before putting your motor in gear, make sure that you are aimed in the right direction.

SELECTED READER ACTIVITIES

1. Examine your inventory of personal skills and abilities, and determine your strengths and weaknesses in dealing with conflict.
2. Engage in a discussion with someone you trust regarding your personal reputation for being able to deal with conflict.
3. Visit the bookstore and obtain a few books on negotiation or conflict resolution skills for your personal library.
4. Prepare a list of activities that you believe could compromise a fire chief's credibility if they were to become public knowledge. Compare this list to the job descriptions in the job flyers to see whether any of these activities are clearly indicated as job requirements.
5. Review your personal background for incidents that tested either your conflict resolution skills or your ethical considerations.

LEARNING DEPARTMENT HISTORY

CHAPTER 10

Cleaning Up Your Act and Clearing Out the Closet:

Examining the Department

The objective of this chapter is to suggest some techniques a new chief may wish to employ to assess the status of the organization when assuming command. The emphasis in this chapter is on professional inquiries that lead a new chief to either confirm the current practices as appropriate or set new directions for the organization based on inconsistencies that emerge in this process. This is not the same as an audit, but it does encompass conducting a systematic inquiry to make sure that the organization is credible and practicing what it preaches.

> *There are two methods of acquiring knowledge, namely by reasoning and experience. Reasoning draws a conclusion and makes us grant the conclusion, but does not make the conclusion certain, nor does it remove doubt so that the mind may rest on the intuition of truth, unless the mind discovers it by the path of experience.*
> —Roger Bacon (1220–1292)

FILLING SHOES AND MATCHING STRIDES

"The King is dead. Long live the King." That exclamation was once used when a king passed away. The old king was dead, but the new king was alive. The transition of power in monarchies was usually swift. Seldom was there any hiatus, for that only created confusion among the noble class, and often resulted in civil war between siblings.

There is a parallel in the fire service when one fire chief leaves, for whatever reason, and another one takes over. When a fire chief vacates a position, one of two things happens. Either someone is placed in an acting capacity to ensure that there is no void, or an appointment is made right away for continuity purposes.

If you happen to be the person selected to pick up the reins, there are consequences, and you should be aware of them. In almost all cases, there are both threats and opportunities during the transition period; depending upon how you handle each situation, you can either benefit or suffer from the outcome. Two distinct areas demand your attention: (1) the impact of your predecessor's administration, and (2) your own biases and your potential for influencing the department by your immediate actions.

You need to be aware of the impact of your predecessor because that person has left fingerprints on the organization. Awareness of what specific actions were taken by your predecessor, what that chief stood for, and did to implement specific ideas or philosophies can be an important part of managing the transition.

At many retirement dinners, I have heard exclamations about how hard it is going to be for the next person to fill the shoes of the chief. That can be true, even if the chief was totally incompetent, since the shoe size of that individual may be entirely different from that of the person who will be assuming the responsibility for leading the organization. But shoe size means nothing; stride means everything. Shoe size is the footprint that you put on the ground, but stride is what a person does in the way of maintaining the momentum in direction of the organization. You can be in someone's shoes and stand still, but if you want to match strides, you have to move.

One of the most important things you can do at the outset of assuming the role of chief is to take a sufficient amount of time to evaluate the performance of your predecessor. What did that person accomplish over the years? What were the apparent goals and objectives? What types of activities, events, and decisions were linked to the chief's management and leadership style? In reference to stride, how much was actually accomplished by that chief, and over what period of time? What were the specific factors that led to his successes and failures?

There are a host of scenarios that will emerge from this type of analysis, but basically there are three types of chiefs that you can supersede. The first of these is the average chief. The average chief does an adequate job, does not suffer any negative feedback, and is not in the midst of any particular controversy. The second type of chief is a legend in his own time. This is the type of chief who appears to be larger than life to the department and to the community he serves. The last type of chief is the chief in trouble. This type gets out of town just before the tarpot and feathers show up.

Looking at the circumstances of your predecessor will help you to determine which activities you might want to implement in matching your predecessor's stride. For example, if you follow a chief who did little more in the department than maintain the status quo, the entire organization may have a sense of ambivalence about change. If you try to move too rapidly, it could create culture shock for the organization. The chief who has been a legend in his own time may be difficult to measure up to at times; therefore, you will want to be very selective in the activities that you try. Following individuals who have reputations of that nature often requires that you make your own mark for different reasons in the community. And finally, if you follow a chief who was in trouble most of the time, you will need to find out specifically why the chief was in trouble. There may be systemic problems or personality problems, or there may be a combination of different types of problems.

Awareness of the previous chief does not mean deference to a previous style, nor does it mean delay in exercising your own style. It just means that you take the time to develop a profile of what worked and what did not work for your predecessor before you set our on your own agenda.

That brings us to the second point, awareness of *yourself.* In the euphoria of being given a chief's job, individuals often lose sight of the fact that they have both strengths and weaknesses. These attributes can either aid them or be liabilities. When you receive an appointment, you need to sit down immediately and take stock of who you are, what you believe in, and what your capacities really are. Your ego may be stroked by the appointment to the fire chief's role, but the danger of derailment can start at the same time.

Confidence can create arrogance. Self-awareness does not mean self-criticism. What I am suggesting is that you take the time to reflect upon the knowledge, experience, skills, and abilities that have brought you to the place of appointment to the chief's role. If you are honest with yourself, you will begin to realize that you are like everyone else. You have both assets and liabilities as an individual. You need to know these well if you are to have any advantage over your adversaries.

There are many things that you will be able to do because you have the vision and the energy to do them. There are many things that you will call upon people in your organization to do because you do not have the skill or knowledge or ability to do them yourself. You need to assess your ability to co-opt others to continue with an agenda that you have decided upon. Failure to do so could result in your ultimate isolation as a fire chief.

Respect the process of transitioning from a former administration into yours, and take into consideration the impact on the organization itself. Because of what your predecessor did and what you are going to do, there will be changes. In many cases, you are not the first chief, nor will you be the last chief of the organization. People in the department will have many assumptions, expectations, and preconceived opinions of what you are capable of doing for them. Analyzing your predecessor and comparing his actions with your own perspective might give you a leg up on what you need to do during the first few months on the job.

The new husband of a woman who had been married almost a dozen times, when asked how he felt about having to live up to his bride's expectations, replied, "Look, I know exactly what I must do. The only thing that I am concerned about is how to make it interesting!" As you enter into a new relationship with an organization, everybody knows what is going to have to happen. The question is whether you will be subtle and sophisticated enough to make it both interesting and successful.

FOLKLORE, FANTASY, AND FICTION

As if things are not complicated enough in our contemporary society, sometimes people start rumors just to add to the dilemma. Have you heard the one about the individual who supposedly stole a cactus out of the desert?

Reportedly, he took it home and put it in his house, unaware of the fact that this giant cactus contained thousands of spiders and centipedes. Supposedly, the cactus fell over, rupturing and discharging these venomous little creatures into the man's house, which he then had to tear down and destroy in order to protect himself. I do not believe this story; it is one of those urban myths that is repeated at cocktail parties and over backyard fences. These myths even appear in newspapers from time to time.

An interesting characteristic of folklore is that it often has a basis in fact. If you went out into the desert and looked long enough and hard enough, I am sure you could find a cactus that contains a lot of venomous little creatures. Whether anybody would be stupid enough to put it in the back of a car, haul it home, and put it inside a house is another question. Yet, the fact remains that plants and wildlife are symbiotic, and from that truth the legend is born.

When it comes to evaluating urban folklore, one of the things that gives it a certain degree of credibility is the fact that it is possible. These urban myths are close enough to things that have happened to other people that, even if we doubt the truth of some of the more outrageous myths, we often respond to them with laughter and the statement, "Well, it could happen."

In many organizations, there is something that parallels the urban myth. I call it *management folklore*. Management folklore is created by the continuous telling of the same story over and over again about why things are a certain way, when in fact that is not the reason they are that way at all. Management folklore consists of explanations that have been repeatedly offered in the past without anyone ever going back and finding the root cause for the position in the first place.

The real problem with management folklore is that the explanation given for the way we do things is not accurate, but it is at least partially true. It may be an acceptable explanation of why a practice was done in the past, but in many cases, it is completely inaccurate with respect to the reason for continuing the practice in the present. The consequence of management folklore is that acceptance of a mythological reason for a practice often leads an organization to perpetuate practices that are outmoded, archaic, or even counterproductive. You need to find these myths as quickly as you can.

Fear of challenging management folklore results in acceptance of invalid explanations for costly behaviors. For this reason, individuals who are responsible for bringing about change in an organization must focus on the management folklore from time to time in order to bring some rationale back to the process. One of the best times to achieve this goal is during transition.

Imagine that something your department is doing is under scrutiny from the outside world. You sit down with your staff and ask, "Why are we doing this?" The response you receive in this situation is likely to begin, "Because . . ." Someone who says, "because" is making a statement about what he believes to be the initiating reason for a particular policy or previously made decision. *Because* means that the cause will follow, which is a trap in itself, since the person will not go back to the point in time that the practice was initiated but, rather, will provide you with an explanation of everything that has been done to support it since its initiation. In other words, instead of giving you the reason for the decision, the person will give you the consequences of its implementation, tell you how long it has been a practice, and may even pepper the explanation with anecdotes about times when the particular item was challenged but upheld. Often the person will fail to tell you the reason the decision is relevant today.

This is exactly how folklore gets its start. When we perpetuate stories about something rather than going back to look at the reason for its existence, we shroud the origins in the mystery of time. And then, over time, the original reason becomes modified by experience and facts or becomes institutionalized in the department to the point where the answer to why something exists is simply "just because."

As a result of having to sort my way through a lot of these management myths, I have adopted a policy of restating the question when I want to thoroughly understand the reason for a policy or practice. Instead of asking, why we are doing something, I now ask, "What was the original reason for this policy [decision, procedure]?" Although this may seem like a subtle play on words, the implications are very specific. When we ask why, we are looking primarily for motives, but when we ask for reasons, we are looking for logic.

If you explore most of the rules and regulations contained within fire department documents, you will find that most of them came into existence "because" somebody did something that someone else, usually the chief, disagreed with, was upset by, or disliked. That person's reaction, however, is not the reason for the rule. The reason for the rule is simply to stop a behavior, require a behavior, limit a behavior, or set out some parameter that is to be followed by everyone. Rules are supposed to require personnel to exhibit certain types of behaviors that are in the interest of the organization. In other words, somebody can do something that causes a rule to come into existence, but the rule is put there to make sure that the specific event never occurs in the future. We need to be careful to separate cause from rationale, especially when it comes to providing written documentation and guidelines for the behavior of our personnel.

Almost every fire agency has management folklore embedded within it. Not all management folklore is bad. Some of it is very accurate. It is a good idea to remember why we do things. The key is to constantly remind ourselves that if the reason remains current, then the rationale should be contemporary. The danger inherent in folklore is that we may perpetuate a behavior within an organization when the logic is not there to support it.

One of the more interesting aspects of folklore is the fact that it actually transcends generations. It is not uncommon for folklore to begin with one generation, be handed off to another, and then be perpetuated by a third generation without any awareness that this process has occurred. Strong-willed individuals in the early days of organizational development often put into play requirements that are carried on by their successors out of respect and admiration for the past. It becomes interesting when a generation eventually tries to go backward in time to find a connection, only to realize that the perpetuators are either dead, retired, or unavailable, and no one knows *why*. What has left the organization with the deaths and retirement of its members is its institutional memory.

And that is the bottom line of management folklore: the more that an organization focuses on developing a written institutional memory, the less likely it is that management folklore will become an aberration. When specific methodologies used by an organization are written down, archived and available for retrieval, management folklore tends to be less fantasy and more a part of the fabric of the organization. But when an organization does not document, archive, and make this information available for retrieval, folklore takes on a connotation of myth and magic. In the worst case scenario, a department not only is unable to explain why it does a lot of things, but also has lost sight of its own roots.

Legacies and legends are all part of the oral tradition of organizations. We should prize our legacies and continue to perpetuate anecdotes and stories about why things are the way they are. In earlier societies, the storyteller was a man or woman of great wisdom who possessed memory of a long line of folklore. In contemporary society, however, we have few skilled storytellers, since we rely upon computer memory, file cabinets, and technology to tell our tales. Personal remembrance is as important as having information documented, but oral tradition can have huge gaps.

The next time you hear someone mention a policy or practice, the reasons for which are lost in the obscurity of the past, ask yourself, "What was the original reason for doing that?" If you can arrive at a rational explanation for doing it in modern times, then the management folklore is an asset. But if the only justification for continuing the practice is because that is the way it has always been done, then, like the dragon slayers in the fairy tales, you should go slay the dragon.

FORMULA FOR FAILURE

Probably one of the quickest ways to program your own failure as a fire chief when you come in from the outside to a new organization is to try to clone your old organization. Trying to force the new organization to fit into the mold of the organization from which you came is actually making two mistakes at once. First, you are taking a lazy approach to managing the organization by trying to merely recycle past practices and policies. Second, you are assuming that there is something wrong with the new organization.

Time and time again I have seen people come into an organization and instantly start the process of cleaning house. They have begun to discard policies and procedures that have long-term tenure in an organization and substitute policies and procedures with which the new chief is familiar. One individual even went so far as to adopt a policy manual for a new department that contained the letterhead from his previous organization. Taking such actions before engaging in a complete analysis of an organization is downright foolish. It creates suspicion and doubt on the part of subordinates about your having any level of trust in the organization. Further, it may foster a great deal of resentment toward the influence of your former organization.

When I made a change from one department to another, my arrival was preceded by rumors of how I was going to make my new department into a clone of my old one. In my former department, I had yellow fire apparatus. One rumor was that I planned on turning the red fire apparatus in the new department into the saffron yellow of the former organization. This rumor was perpetuated without taking into consideration the fact that the fire trucks were painted yellow in my former department for a very specific reason. That reason did not exist in the context of the new organization.

There is a reason for doing most things. As we change organizations, the reasons for doing things often change. There is an inherent danger in assuming that because something worked once, it will work in a new context. To the contrary, as the old cliché goes, "When in Rome, do as the Romans do."

I am not implying that you should roll over and play dead in an organization that is not going anywhere and needs to be given direction. But you need to make sure that you separate real change from mere cloning because of familiarity. As the leader of a new organization you have an obligation to make sure that the changes you introduce are meaningful, relevant, and appropriate. Imposing past practices on an organization without completing some form of review process is an extremely superficial behavior that may, in fact, increase your vulnerability to criticism as a decision maker. The temptations are great. If something was once a success, we are tempted to try it again and again. You must be disciplined to make sure this temptation does not take the place of a thought process.

There is a similar concern for individuals who move up within an organization. Depending upon the relationship between the old chief and the new chief, there may be a narrow or an extensive gap between the perspectives on how things should be done. To recoin an old cliché, "A new broom might sweep clean, but it can also sweep out the good with the bad." An overly enthusiastic leap forward can create both backlash and black holes. In the case of the former, a newly promoted chief from the inside tries to move the organization faster than it is prepared to move. In the case of the latter, too many changes made too fast can create gaps in the organizational culture. Both are traps for people who just can not wait to charge in and change all the things that have frustrated them in the past.

One thing you should keep in mind is the difference between preferences and policies. Almost all of us have some personal preferences in regard to how we would like people to interact with us and how we would like to interact with them. Policies, on the other hand, are very generic and apply to an entire organization. Whenever you begin to replace policies with your own personal preferences, you increase the possibility that your personal preferences may not meet the general needs of the organization.

You may have created policies in your previous organization or rank that were based on your experiences in that position, and you may have developed personal preferences. In the context of organizational transition, however, changing a policy based on previous acceptance of your leadership does not ensure that the new policy will be accepted by the individuals in the organization. To the contrary, they will be fearful and potentially threatened by this set of circumstances. In one sense, you are intimating to them that they have been wrong all along. Too much of this behavior will begin to convince your personnel that you have no faith in them or their organization, and you will begin to get a gap between you and your subordinates.

Let me give you an example from one of my transitions. I have some preferences with regard to how I like to interact with my command officers. I have some styles of communicating that involve regularly scheduled meetings, agenda management, and personal accountability. When taking over the organization, I spent a considerable amount of time discussing these preferences on a one-to-one basis with the command officers. Therefore, they were startled that the first command I put into place in the organization was that all policies, practices, and performance criteria of the previous administration would remain in full force and effect until such time as they were modified through a decision-making process at the staff level. Important controversial items, such as hose load configuration, tactics and strategy, and organizational and program activities were to remain at the same level as they were at previously. Some of the chief officers had been operating under the assumption that I was going to modify our procedures right away to make them look more like those of my previous department. The publication of the special order announcing this stabilization was an important element in stemming that rumor.

Does all of this mean that it is impossible to bring about change in an organization during transition? Absolutely not! But you need to make sure that when you introduce change, you do so in such a fashion that it fits the context and does not become counterproductive. There are specific techniques you should avoid when you introduce your ideas. Probably the most obvious one is that you should never refer to practices conducted in a previous organization. "What we used to do back when I was a member of . . ." borders on offensive to individuals who were not in that previous organization alongside you. Most people have a reasonable sense of pride in their organization, and they do not like to be told that while they were working to obtain their knowledge and experience, someone else was doing it much better. In the 1960s, many fire department offices had signs that read, "We don't give a damn how they do it in Los Angeles." The fact is that most people really do not care how something is done somewhere else. What they do care about is how they do it and how well they do it.

The second suggestion for making changes without falling prey to the cloning syndrome is to make sure that you do your homework before you institute any significant policy shifts. Earlier we talked about conducting a self-assessment. You also need to make sure that you understand all of the assumptions that bear on the context and culture of the organization. This will do more to maintain your credibility than you might believe. Someone once said that if you want to make people like you, do not talk about yourself, but ask them to talk about themselves. When you are going through this transition period, spend time with the members of the existing organization, finding out their likes and dislikes. This might reveal to you an interesting surprise—they might want to do exactly what you want to do.

Another step in preventing backlash is what I call the triangle of educate, incubate, implement. If you have some practices that you would like to see brought into the forefront, the best way to start that process is to provide staff and members of the organization with an opportunity to receive information about the practices through the educational process. This can be quite time-consuming. It may involve anything from arranging for educational opportunities to give people a chance to learn about something from others to the more subtle technique of merely providing an adequate library and resource materials in the fire station so that people can study on their own.

The incubation part of this process implies that there is a waiting period between ideation and implementation. It is not uncommon for this part of the process to result in individuals coming to you with suggestions to make changes that are exactly what you want. The changes will sound a lot better to the members of the organization, however, when the ideas come from them.

The implementation phase is obviously the time for action. The lower this implementation can be delegated down in the organization, the better. Trying to make changes yourself implies a personal involvement that, in some cases, may be too intimidating for subordinates to accept. For example, it is not appropriate for a fire chief to dabble with changes that need to be made at the fire company level. This does not mean that you have no interest in those changes; but if you are the one making the changes at the company officer level, then you really do not have any need for your battalion chiefs or intermediate command officers. The implementation phase usually takes a lot longer than we would like for it to take. That is when our patience must come into play.

The last suggestion regarding the transition of a department is to maximize the opportunity for people to have input into the change process. Upon taking command of one organization, I discovered that the policies and procedures manual was essentially a good manual but had not been properly maintained. I determined that this would be an excellent opportunity to acquire support for some needed changes. Therefore, rather than discarding the manual and creating a new one, I appointed a task force of senior officers and company officers in the organization to conduct a review of all existing policies and procedures. The committee was given the responsibility of reviewing the current manual in order to make three determinations. Was the policy currently being followed? If not, how did the policy have to be revised in order to make it congruent with the department's operations? Should the policy or procedure be maintained in the manual or discarded in its entirety?

Before they began their review of the policy manual, several of the command officers were given an opportunity to attend state and national fire academy courses dealing with administrative procedures. One individual had the opportunity to visit a series of neighboring fire departments to determine what policies and procedures were in their manuals. The result was that it took over a year to rewrite the manual. There was a considerable amount of debate at the officer level about the answers to all three questions. The net result, however, was that when the manual was reissued, it was realistic in regard to what was going on in the organization. Interestingly enough, although they had not been excited about the project initially, the members of the task force took it upon themselves to make suggestions about policies that needed to be added to the document.

During this transition period, I also had the opportunity, as chief, to engage in conversations with the command officers. I asked a number of questions about policy changes that I felt were needed so that, during the committee activity, these issues got addressed. When the new policy manual emerged, it was much better than the previous one. It contained some components of my previous experience in another department, and yet, it was unique in its own right. It was built upon the needs of the department that would be using it.

Some of the techniques discussed here will require a great deal of patience on your part. You have to be prepared to finesse solutions instead of forcing them. When taking responsibility for a new organization, remember that significant change in an organization cannot be brought about too quickly without causing some kind of harm. The person responsible for introducing, adopting, and molding change in the organization is the chief. If you are cast in the mold of merely trying to duplicate your previous organization or to adapt the organization to fit with your personal biases, people will feed you information that is designed to placate that desire, but in some ways, this will become counterproductive. You do not want a clone; you want an original. What you want is to reach the greatest potential of the organization, which means that, in many ways, your new organization may be better than the last one.

CITSA: ACRONYM FOR THE FUTURE

The fire service has a penchant for using acronyms. Because we are a technical profession with complex techniques for addressing specific issues, we use acronyms as a form of verbal shorthand to convey concepts quickly. An acronym uses as few letters as possible to convey as large a concept as possible in an efficient manner. There are simple acronyms, such as rpm and psi, that we use even on the fire ground. And there are more complicated ones that we use in staff meetings to discuss budgets or computers, like MBO and WYSIWYG (management by objectives, and what you see is what you get).

There is a new acronym that is emerging from the widespread interest in making the fire service as cost-effective and efficient as possible. It is CITSA, which stands for constant improvement through self-assessment. Although it is not exactly firehouse slang yet, its future seems to be assured. The reason is that fire services all over the world are coming under the scrutiny of political questions regarding costs and benefits for public services.

There are several forces driving this phenomenon. One of the most obvious is the increased emphasis upon the total quality management concept used mostly in the private sector. Another impulse is created by the state of public finance in most areas where there are more demands for service than there are funds to provide them.

The purpose of this section is to focus upon a method to assist a newly appointed fire official in meeting the challenges posed by the previous sections. The process is called the Fire and Emergency Services Self-Assessment Process. Some people are labeling this process as the "accreditation" process, but that is not accurate. Self-assessment is a process. Accreditation is an outcome. They are related, but they are not the same thing.

The purpose of this section is to zero in on how self-assessment is a tool that can and should be used to respond to the inquiries of policy makers regarding what a fire agency means to a community. How can a fire agency prove that it is doing what it should be doing to protect a community, and how can it guarantee that it can sustain that effort? How can this technique help you to get a handle on what is going on in the organization during the transition period?

Once upon a time, we all looked upon the person who had all the answers as the expert. The world has changed. Today, the person who is most often looked to for guidance is the person who knows the right questions to ask. In the context of a fire agency, that person should be the fire chief. This phenomenon is based upon an amazingly simple fact. The answers keep changing in our society so fast that yesterday's answer may be inaccurate today. This fact provides an opportunity for the self-assessment process to become a meaningful tool in the hands of a fire official.

Here is how it works. Self-assessment consists of creating an internal audit process, using a structured sequence and pattern of questions that are not just sequential, but synergistic. Self-assessment is a methodology that requires two things to occur simultaneously. The basic questions to be answered are: What are you doing? Do you know why you do the things you are doing? Is the way you are doing things the best way?

The manual I refer to in the following text as "the self-assessment manual" is the *Fire and Emergency Services Self-Assessment Handbook,"* published by the Commission on Fire Accreditation International (CFAI) in Fairfax, Virginia. A person does not need the manual to understand the concept, but if you are interested in pursuing an accreditation outcome, you will need to purchase a copy from CFAI.

The self-assessment manual proposes the following: the creation of a self-assessment team, the writing of a self-assessment document, and the application of those findings to bringing about improvements in the agency over time. A separate decision is required to take the self-assessment process into the accreditation program, which involves payment of fees and the requirement that an outside group of peer assessors visit the organization to validate the credibility of the self-assessment findings. I am not suggesting that the full accreditation process is needed by a chief during transition; but I am suggesting that the process of self-assessment can be a powerful tool to assist in clarification of the present conditions and the future needs of the department. Any fire agency can engage in the self-assessment activity. It is an excellent tool to deal with the need to justify, articulate, and refine an agency's role.

There are at least three times in a fire official's career when the self-assessment process is particularly useful: (1) upon initial appointment as a fire official, (2) during a period of major changes in the organization, and (3) when a person is about to turn the organization over to a successor. The process works just about the same whether you are in a take-over, make-over, or hand-over mode. The differences that will emerge at different times are dependent upon the maturity of the organization. In this chapter, we are focusing on the take-over mode.

The self-assessment process is not described here in detail in terms of its size and level of participation, but the basic concept is simple. The idea is to take a long, hard look at what a fire organization is doing and to make some expert recommendations on how things *could* be done better. In order to accomplish this, a group of individuals, a self-assessment team, takes a critical look at the answers to the series of questions raised by the self-assessment process.

The first step is to determine which individuals in your organization are "critical thinkers." (I am using the word *critical* here to mean in-depth, as opposed to cynical or jaded.) In some departments, the team may be limited to one or two top managers, while in others, it may be a widespread resource. The self-assessment team needs to be created from those who possess the ability to be both objective and professionally curious at the same time. The former quality is required for fact finding, and the latter for innovation. Both qualities are needed in the team members to make self-assessment result in improvements. Creating a self-assessment team during transition helps to clarify future team relationships as well.

The self-assessment process contains a simple concept of drafting an objective collection of facts and the documentation of improvements that should be considered. The self-assessment system requires that when a response is prepared to the structured inquiries, it is divided into three parts: a description, an appraisal, and a plan. In essence, this format requires that all responses contain facts, opinions, and recommendations. A description is the collection of facts that can be verified. The appraised position requires the expression of an opinion about whether the facts are appropriately configured to address the needed performance in the organization. The plan requires documentation of what adjustments in current policy, procedure, or practice could make the organization function better.

The body of the self-assessment process consists of a structured set of inquiries. They are divided into ten categories of criterion and performance indicators. The ten categories are not just a random listing of areas of inquiry, but are structured to be interrelated. The categories are:

1. Governance and administration
2. Risk assessment
3. Goals and objectives (management methodology)
4. Finance
5. Programs
6. Human resources
7. Physical resources
8. Training and competency
9. Internal relationships
10. External relationships

Here is how these categories interrelate. The system assumes that a fire agency must be given the authority to act to solve a specific mission for any agency; this mission must be based on what is to be protected; and everything must flow from these two elements to be legal and to be effective. This leads to the need to have an appropriate direction through goals and objectives that resolve (reduce, contain, or modify) the statutory responsibility. This leads to an assessment of proper funding support and prioritization of an array of programs to resolve the problem. Funding and programs are linked not just to the cost, but also to the benefits. The agency must address its human resources and physical resources to achieve the level of performance described in the program, which is linked to the first two categories. Training and competency is a bridge requirement that focuses upon the need of the organization to train in order to protect what has been defined by the analysis of the first six items. Lastly, internal and external assessments need to be made to see whether you can sustain the effort described so far and whether the organization is operating in concert with a broader context or in isolation.

The criteria are merely subsections or steps in each category used to demonstrate that the organization is pursuing the category in a comprehensive manner. The performance indicators are proof that an activity is underway and can be defined, described, documented, and reviewed.

The embodiment of a self-assessment process is an actual report, written by members of the self-assessment team and edited by the leaders or facilitators of that team to ensure consistency in format and styles. The self-assessment manual, which is essentially a word processing document, contains only narrative. It does not contain examples or exhibits of actual documents. These are collected and used to verify the contents of the self-assessment manual.

When the self-assessment manual is complete, it has at least three uses. Every response that describes a criterion or performance indicator is useful as part of the day-to-day assessment process of the department. This portion has often been referred to as a "desk reference." The second use of the manual is that the appraisal portion of the manual often contains information on gaps in the organization's actions that can be remedied in either the short or long term. The short-term items can become part of the tactical or work plan and quick fixes that a new chief can adopt to make meaningful first steps in the agency. The last part of the manual, the recommendations, can serve as a long-term inventory of things to be done and as the basis for a strategic plan. With computer word processing capabilities, the manual can be easily edited into three work products. Some agencies choose to use the output as a whole, but three separate modes are possible.

Self-assessment is structured so that you have a couple of options available for using the information. The first and most obvious one is to take what you now know about the organization and recycle that knowledge into budget proposals and public information. I say this is obvious, but in reality, many departments make so many assumptions about the basic information that this area is, perhaps, one of the fire service's greatest weaknesses. Sometimes political entities completely forget what tasks they have charged agencies to perform. The community hazards, risks, and values often change so gradually over time that both the agency and the political leadership forget what the consequences of inadequate planning place at risk. Self-assessment provides data that are important to the clarification of a fire agency's reason for existence.

The second option for using the self-assessment is to use the elements to direct and redirect the priorities of the department. Take note, however, that self-assessment is not intended to promote wholesale revisions to a department. To the contrary, to paraphrase W. Edwards Demming, incremental improvement is to be preferred over the paralysis of analysis.

In conducting an evaluation of the self-assessment process, I have observed that the creation of the base document takes about one thousand person hours. This figure means that two people, for example, could do it in about five hundred hours each, while four would need two hundred and fifty hours each, and so on. On the surface, that sounds horrific, but is it? A person has about fifty workweeks per year. If one person devoted only ten hours per week, that would be five hundred hours per year. This speaks to the issue of how many people belong on the team—the more the merrier, but, more importantly, the more the quicker it gets done.

This leads to the maintenance of the document, which is a fraction of the time needed to create it initially. CITSA depends more on routine review of the process than it does on the initial draft; therefore, to maximize the use of the information, someone, or everyone for that matter, should go back to the notebook and define, refine, and redefine improvements as they materialize.

There are no perfect fire agencies. All agencies can be improved. But we must face the inevitable: sooner or later, improvements become more and more expensive to accommodate. I am proposing here that a new chief use self-assessment not as a finished product but, rather, as a tool to get a solid foundation for future management decisions.

What CITSA does is to place emphasis on managing change so that the organization becomes more credible, and therefore, less vulnerable to unreasonable and irrelevant attacks upon its basic framework. It will not impart wisdom to new fire chiefs, but it will give them a solid foundation of facts. It will not create the ability to overcome other priorities that are well-defined and appropriate competition for the same resource dollars, but it may level the playing field.

ORGANIZATIONAL ARCHEOLOGY: CLEANING OUT THE CLOSET

In the late 1940s, there was a famous radio show called "Fibber McGee and Molly."[1] One scene that was part of almost every show was one in which a closet was opened to retrieve some meaningless object followed by the loud sound of clatter and clanging as debris rained from the closet onto the heads of those who opened it. The hero of the show, Fibber, would always say something like "Someday I'm just going to have to clean out that closet."

---

[1] "Fibber McGee and Molly," National Broadcasting Company, 1935–1959.

All of us accumulate items in our day-to-day lives that we retain out of fear that we will need them someday. I am a lifelong collector; I have literally hundreds of thousands of items going all the way back to my childhood days in school notebooks. In the context of our fire departments, however, the accumulation of this kind of material takes on a different connotation.

I am referring specifically to the record-keeping system in fire departments. It is among the most important items for you to inventory quickly. Record-keeping systems are somewhat like the closets of our homes. We tend to accumulate objects that are placed out of sight and out of mind. In the fire department, statistics are among the things that we lose quickly. Accumulation without evaluation ultimately results in a massive collection of basically meaningless information. This is not to say that record keeping is unimportant to the fire department; it is absolutely essential. However, keeping records for decision-making purposes is somewhat different than keeping records, period. In working with fire departments, I have discovered that, in many cases, record-keeping systems started out with relatively noble purposes but eventually accumulated a wide variety of reputations, most of them bad.

In a lot of cases, the information being collected by fire-fighting agencies is used for defensive purposes only. In other words, the information is collected only for the purpose of making sure that if someone asks the question someday, the answer may be available. In other cases, records have been kept because they were needed at one time, but when the reason for their existence disappeared, the records were retained. Those who recall David Gratz's text, *Fire Department Management: Scope and Method,* will recall Gratz's discovery of a record-keeping system that was tracking serial numbers on tires. It was originally created to keep track of tires for the war effort, but in the 1960s, in spite of the fact that World War II had been over for twenty years, the system still existed.

We tend to take our record-keeping systems for granted. They are there because they were created at some point in time and seldom do we take the time to evaluate the need for continuing them. Instead, we often add layer upon layer of records on top of an existing system until it becomes difficult to determine the purpose of any given document. As a matter of fact, in many cases, documentation clearly illustrates that there is redundancy of information and counterproductive activity in record-keeping systems.

The solution to dealing with this problem is called a "records audit." It is a simple, straightforward process that is utilized infrequently in the fire service, probably because it requires looking at something that, in a lot of cases, is fairly ill defined. There is a certain degree of fear associated with examining that which is poorly defined and, at the same time, traditional.

Doing so during a transition period is less threatening and may provide more insight than doing so at any other time.

The records audit consists essentially of the following elements:

A. A documentation of all the existing forms, records, and record-keeping systems utilized within the department

B. An assessment of the utility of each of these documents in the modern setting

C. The decision-making process used to combine, revise, eliminate, or reestablish specific documents

D. Establishment of a records retention and revision timetable to prevent the cluttering of records in the future.

The documentation of an existing records system is a lot simpler than it may appear. In its most simplistic form, it consists of taking all the forms that are in existence in a fire department, putting them in some form of organizational structure, and documenting the various elements of their use. It is important that a record exists within the context of an overall management information system. Therefore, all records should have some type of form number. The form should be clearly identified by title, and it should be identified with respect to who initiates it, who utilizes it, where it is stored, and how it fits into the overall scheme of the management information system.

The second element of the records audit, regarding the utility of a form, is much more difficult to address. Almost all records consist of a variety of data elements, or facts, that need to be reviewed in the establishment of that form. Each document has to have some form of address to demonstrate where that document fits into the system as a whole. It is extremely important, in performing a records audit, that each and every document that is utilized by the department be evaluated in regard to its contribution to a decision-making process of the organization.

Frankly, there are many documents utilized by fire departments that are kept for legal reasons only. In many cases, records came into existence because a problem existed and someone decided to start keeping records on that specific issue. The determination of the utility of the document rests at essentially three levels. The first level is the corporate one, that is, the city or municipality that is asking the fire service to provide a certain function. If they do not need the information, then the question is, who does? The second level is the managerial level of the organization. This level involves the fire chief and the key staff. If they do not need the information in order to manage the department, then is the record necessary? The third level is the operational level, which, in many cases, is the determining factor in whether a document really contributes to the overall effectiveness of the organization.

Record-keeping systems should be viewed somewhat like pyramids. That is to say, most of the data is collected at the lowest possible level in the organization, and as it moves upward, it accumulates in significance and begins to identify trends and patterns of the organization. A classic example of this might be the training record. The individual training records of an individual fire company on a daily basis are relatively meaningless. When that information is accumulated on a monthly basis, however, it tends to paint a picture of the overall productivity of the training program. Collected on an annual basis, it may give a direct correlation with respect to the individual's exposure to specific techniques or procedures. I know for a fact that in one case, this type of information was extremely relevant to a court case involving an apparatus collision in which there were fatalities. The attorney for the plaintiff petitioned and received information regarding how much training an apparatus operator had received in being prepared to function in that job.

The accumulation of data in this pyramid-like manner means that as the data moves from the basic level to the operational level to the managerial level and subsequently to the corporate level, it must become increasingly more simplified, or it needs to be characterized differently. For example, in the case of incident records, nothing would be more startling to the city manager or mayor of a community than to receive an entire stack of incident reports from the fire department for a given month or year. Summaries of data are meaningful only when they indicate trends and patterns or measure overall effectiveness or outputs of an organization.

Therein lies the necessity for a records audit. Quite frequently, as documentation occurs at the bottom of the pyramid, large amounts of raw data are collected but never used anywhere else in the system. Essentially, the information is collected, collated, and stored away in file cabinets never to serve a purpose. This is a counterproductive activity, since this information is collected at the expense of other information that *could* be collected.

This leads to the third phase of the records audit, that is, the conscious decision on the part of management or operations to make changes in the record-keeping system. Simplification is not merely a case of just cutting elements from a system. Simplification consists of evaluating data elements to eliminate redundancy and to improve the usefulness of the information. A records audit in one organization revealed that the department used more than one hundred and fifty forms to collect different types of information. A comparison of data elements on these forms, however, revealed that twenty to twenty-five of them collected almost identical information but in different formats and on different documents. By creating a combination of these documents, or multiple-use forms, the department eliminated the need for several forms and streamlined the operation.

The purpose of making operational changes in reports is not to throw out elements of the record-keeping system. Instead, the purpose is to have each data element contribute to the managerial- or corporate-level decision-making process. One spinoff of the records audit is that it often points to the justification for automated record-keeping systems, that is, computerization or the use of "batch processing" types of systems to facilitate the collection of information.

The last element of the records audit is the establishment of a records retention and revision timetable that will prevent the "Fibber McGee closet" from occurring again. In the first phase of the records audit, that is, documentation, one decision that needs to be made is with respect to the filing location and the length of time that a document is significant to the organization. One time, in performing an audit of this type, I discovered four file cabinet drawers full of payroll records that went back almost fifteen years. In the process of evaluating those payroll records, it was determined that a duplicate copy had been maintained by the city in its archives. This was an unnecessary volume of documentation, and permission to destroy the records was relatively easy to achieve.

There is a certain degree of paranoia associated with the destruction of a record. Frequently you will face a lot of opposition in cleaning out the closet. There is an implication that if it was important enough to collect in the first place, it may be important enough to keep forever. Nothing could be further from the truth. There is a retention period for any piece of information to be relevant. City attorneys in most areas, will gladly share with you what the retention period is for most legal documents. In other cases, for example, apparatus records, there are reasons for long-term retention. In any case, the decision to establish a retention period on all records is a key management responsibility. Failure to do so will result in a massive accumulation of information that cannot be used without exhausting staff energy and patience in finding the information.

One question remains: when should you perform a records audit? If you have never performed one, the answer is right now. If you have performed one and you have followed through with the four steps, there is a high degree of possibility that your record-keeping system is self-monitoring at this point and that a records audit will not need to occur again. The records audit is cathartic. Note, however, that it must be performed with care. The records audit has caused a great deal of concern in organizations where valuable information was disposed of arbitrarily.

The advantage to cleaning out a closet is that once the items in the closet are straightened, you can look into the closet and find anything you need in a hurry. The same thing applies to a records audit. Once it has been performed, one can look at the records management system in the organization and find what one needs quickly and with confidence.

Spring house cleaning is a pain, but it can also be somewhat of a delight. As you rummage through the closet, you often will be surprised to find "real pearls." Copies of fire reports that reflect upon the history of the organization, photographs of fires that have long been forgotten, commendations and historical events that have been obscured by massive amounts of paperwork often emerge from this process unscathed. Quality is much more important than quantity. In looking at the historical perspective of most organizations, the record-keeping systems, as a whole, do not tell a story automatically.

Like archeologists who must wade through layers and layers of dirt, the records auditor has to wade through layers of information to find something that is meaningful enough to be retained. But, like the archeologist who discovers a priceless treasure under layers of mud and clay, the records auditor just may find the one fact that best serves the interests of the fire department.

SELECTED READER ACTIVITIES

1. Write an overview of the previous chief's administration. What was the culture? What were the accomplishments? What was left undone?

2. Examine the management process in the department and ascertain what policies, practices, and procedures are shrouded in mystery. Determine the original reasons for their adoption.

3. Look at the staff of your department and determine who would be the best members of a self-assessment team. What special skills or attributes would they bring to the process?

4. Conduct a review of the management information system and determine if it provides you with the type and quality of information you need to manage the organization.

5. Prepare a prognosis for the department of its future in the next five to ten years. Identify areas that are likely to experience changes.

JUGGLING TIME

DOUG HUBERT PD

CHAPTER 11

Juggling Eggs, Chain Saws, and Hand Grenades:

Managing Your Time

The objective of this chapter is to focus on how a person can get a handle on his capabilities and limitations in order to actually get things done. The purpose of this chapter is to make sure that a chief has a game plan to use his time as effectively and efficiently as possible.

I would rather be ashes than dust
I would rather
   that my spark should burn out in a brilliant blaze
   than it should be stifled by dryrot.
I would rather be a superb meteor,
   every atom of me in magnificent glow,
   than a sleepy and permanent planet.
   The proper function of man is to live, not to exist.
I shall not waste my days in trying to prolong them.
I shall use my time.

—Jack London (1876–1916)

TOOLS, TECHNOLOGY, AND GETTING THE JOB DONE

If you were walking through your garden and stubbed your toe on a rock, you might be slightly irritated. You might kick it out of your way or even pick it up and throw it. It is doubtful that you would give it a second thought. On the other hand, if you were going through a museum and you saw that same rock sitting on a shelf labeled "Tools of Ancient Man," you might give it a second glance.

Rocks were once literally "tools of the trade." A brief visit to any anthropological museum will demonstrate that rocks once were shaped into various tools to be used either as weapons of war or weapons of commerce. The evolution in the use of those tools led to the formation of metal tools and machinery. Ultimately we developed the technology to carry us from the surface of the earth to the surface of the moon.

Technology is a wonderful thing. It is also a double-edged sword. It improves our lives, and yet it complicates them. It makes our lives easier and makes our jobs more technical. Some people have an abject fear of new technology. Others actively seek it. Most of us learn how to cope with it over the period of our lives, although some never do. But what of the future? If we think that a rock shaped into an axe or knife is primitive, imagine how complex our lives must look today compared to the fire service of one hundred years ago.

Almost all fire departments are going to be impacted continually by changes in technology. The manner in which firefighters are aware of these impacts will vary drastically in accordance with the manner in which the organization approaches technology. Essentially, there are four basic philosophical approaches to technology. The people who adopt these approaches can be described as follows: the technophile, the technophobe, the technocrat, and the technomanager.

The technophile is someone who accepts new technology merely because it *is* new technology. You can identify a technophile because the individual is literally surrounded by gadgets. In some cases, the technology fails to meet its original mission and occupies a place on the shelf or in the back room. The technophile has no difficulty in dealing with technology because he accepts it for surface or gratuitous means; in other words, what matters is not whether it works, but whether it is new.

The technophile does have an advantage over the mainstream of society because he is usually on the "leading edge" in regard to the tools that improve productivity or power positioning in organizations. A classic example of this might be in the adoption of miniaturization in the radio systems. Probably the first fire departments that embraced radio technology were those that had technophiles among them who could understand the impact of this electronic gadgetry.

Technophiles love technology for the sake of technology. Their weakness is their lack of discrimination in investing in technology before its potential for making a contribution to their organization has been proven. Probably the world's largest collection of technophiles can be found in the area of camera enthusiasts. Every time a new gadget is invented that supposedly improves photography, the camera buff (or photophile) "simply has to have one."

The opposite of the technophile is the technophobe. This type of individual has an abject fear of new technology. Technophobes reject the technology *because* it is new. A classic statement attributed to individuals who fall into this philosophical orientation might be "if it was good enough for my grandfather, it's good enough for me." In our modern age, few people are willing to admit that they are technophobes. In many cases, individuals who fear technology will fight a silent battle with the new technology by merely making sure that it does not work "for them." The technophobe does not look at whether new technology works, but at what he has to do in order to make it work. If the technology requires a change of behavior on or a reorientation of thought process on his part, then the technophobe tends to reject it.

Technophobia occurred in many offices when businesses and government converted from the typewriter to the word processor. Thousands, if not hundreds of thousands, of extremely competent staff members found all sorts of reasons why the word processor was an inappropriate tool in the office. In spite of the fact that this technology would improve their productivity and, in many cases, reduce the frustrations and anxieties of their positions, they rejected the technology on its surface implications.

If we were to go back in history and examine how office workers functioned before the invention of the typewriter, we would probably find that a similar philosophical rejection occurred when the typewriter was invented. Individuals who transcribed documents by hand probably used the same line of logic to prevent the adoption of the typewriter. Typewriters, however, were purchased, and office workers learned how to use them. And so it was with word processors and computers.

The weakness of the technophobe is the fact that he has no control over his destiny. Rejection of new technology does not mean that it will not become a commonly accepted practice in any given profession; as a matter of fact, the contrary is true. The more outlandish the rejection of a given technology, the more likely it is that the technology will be refined and focused upon until ultimately it becomes acceptable. Witness the proliferation of the flight industry after human beings finally learned how to fly.

The technocrat, the representative of the third orientation, is subtle and difficult to identify. The technocrat is an individual who embraces technology for the express purpose of being able to use the technology as an end unto itself. The watchword of the technocrat is "what's in this thing for me?" The technocrat is not necessarily oriented toward outcome, but more focused on the process of technology. The classic example of the technocrat is the computer programmer. The rarefied language associated with programming can result in a form of technician who is extremely difficult for others to understand and, subsequently, to manage.

The technocrat focuses on process. He can usually provide all sorts of nuts-and-bolts answers regarding specifications, tolerances, and electronic "gee whiz" terminology. The technocrat can justify the cost of any given technology down to the final penny, but in many cases, he cannot tell you what it will do for you, nor can he produce a product with it.

This brings us to the individual with the last philosophical orientation: the technomanager. The technomanager is an individual who looks at technology in the realistic light of three factors:

1. What will it do for me?
2. What is it going to cost me?
3. What will the impact be on the organization as a whole?

The technomanager neither accepts nor rejects technology on its surface merit. Instead, he looks upon technology as a tool—a tool to get the job done, to improve productivity in the organization, or to modify working conditions within the organization. This is a task for the fire chief.

The technomanager rises above the sheer statistical and engineering "tower of babble" that exists within technology. In many cases, the technomanager cannot even begin to tell you how a certain kind of technology functions, but he can tell you exactly what it will be able to do for the organization. Technomanagers grasp generalities and impacts of technology. They focus on that thin line of contact between the tools and the people who must use them. They utilize technology to improve upon performance, both their own and that of others. They balance the differences between how something gets done with the competitive needs of the organization.

Now that we have examined all four of these philosophical orientations, you can probably think of individuals who fit into these four classifications. The technophiles can be extremely expensive for fire service organizations. Gimmicks, gadgets, and trendy items are fodder for technophiles. They collect samples of everything, and their justifications for adopting new technology are often quite superficial.

Technophobes can literally freeze a fire service organization into a status quo orientation that is lethal. Imagine, if you can, an individual who had control over the purchasing power of a fire department in the mid-1920s. Steam fire apparatus was still responding on the streets of our cities, but the automotive fire engine was a reality. The technophobe could have frozen an organization into a technology that was counterproductive for effective fire protection.

Probably the area of fire service with the most technocrats is the organizational structure that supports computer-assisted dispatching systems. Anybody who has ever dealt with the programmers who establish computer-aided design (CAD) realizes that their time and attention are devoted to activities that are somewhat unrelated to the reality of fire fighting in the streets. Communities that have embraced computer technology have often lost their momentum by allowing technocrats to convert the computer system into a bureaucratic nightmare at city hall. Fire chiefs in such communities have found themselves totally isolated from the use of computer technology because the systems have been placed under the direction of technocrats who are reluctant to relinquish their power to hook up to the computer. The same technocrats are also reluctant to provide programming background to the fire service managers to allow them to use the system as a tool on a daily basis.

The fire service has technomanagers as well. All around the country, there is evidence that the concept of technology transfer is taking place within fire departments. Many small, medium-sized, and large fire departments are exploring technological tools to improve their day-to-day operations. In many cases, these tools are not found on the floors of fire ser-vice conventions. Instead, they are found in the yellow pages or in the adver-tising areas of other types of professional magazines. Nonetheless, technomanagers are finding tools to hone the competitive edge of their orga-nizations. These tools include such things as satellite pagers, facsimile trans-mitting machines (fax machines), computer technology, and built-in fire protection.

In the evolution of technology, there is no such thing as the status quo. In spite of the fact that many fire service professionals and members of their organizations believe that there is a "standard," the status quo is an illusion—not unlike a motion picture film that appears to move but actually consists of a series of still frames, each one slightly different from the others. The evo-lutionary process and technology utilization run parallel on a daily basis. The fire service cannot put a hold on technology because it does not control a large enough area of the economy to stop invention from occurring.

Where will it all end? The first response is, nobody knows. The second response is, why would anybody want to stop it? Granted, some technolog-ical advances have endangered lives of society. We fully realize that the hazardous materials the fire service faces today are the results of a techno-logical liability. On the other hand, the quality of life has improved drasti-cally with the development of processes, products, and services directly related to the hazardous materials dilemma. For example, computer tech-nology is being used to fight hazardous materials problems and to assist us on the scene of medical emergencies. It is interesting, however, that the microchip industry used to develop computers is responsible for many of the hazardous materials and medical emergencies that we handle. It is a point-counterpoint situation.

Each of the four philosophical orientations discussed here has its infer-ence for a local fire department, especially when the behavior produced by the orientation is manifested by the boss, the fire chief. The technophilic fire department can become frenetic, ineffective, fragmented, and, in some cases, downright dangerous when it comes to protecting its community. The technophobic fire department can become obsolete, unprepared, and dan-gerous in its own right in protecting its community. The technocratic fire department can become somewhat like a dog chasing its own tail; instead of solving problems, it only describes them in more and more finite detail. The department that is run by a technomanager lives with technology but is not dominated by it.

A fire service manager needs to become a technomanager. The fire chief does not have to become an expert on technology in order to use it effec-tively. Instead, the task for becoming authoritative on a given technology should be delegated to individuals within the ranks who have sufficient training, education, and technological orientation to deal with those specifics. The technomanagers should be asking themselves, "What can this thing do for my department?"

You will notice that, throughout this entire discussion, I have not dealt with the costs of technology. There is a price to pay for everything. Some technology is inexpensive; other technology costs a small fortune. The adoption of a given technology is, in some cases, a managerial decision that rests outside the realm of the fire chief.

The technomanager focuses on value, not cost. An understanding of value is not something that you acquire merely by going to school or having experience. It is a combination of both fact and feeling. As a fire service manager, you will probably find yourself facing technological decisions now or in the immediate future. Keeping technology in its proper perspective will never be easy. But if you look upon it as a management challenge instead of looking at it with fear or anxiety, it can be an exciting decision-making opportunity.

It has been said that there are only two kinds of fools in this world. The first type of fool is someone who rejects something merely because it is new. The second type of fool is someone who accepts something merely because it is old. How about using technology to help you become a better manager?

DO ONE-MINUTE MANAGERS HAVE ENOUGH TIME?

The comedian Robin Williams once had a line that went "Reality—what a concept!" The line used to get a lot of laughs, probably because, for many people, reality and their desires are often totally separate. Williams's line capitalized on the black humor of frustration.

As fire chiefs, most of us have to accept the reality that there are many more demands upon our time than there are minutes in the day. Time management—what a concept! Many years ago Ken Blanchard suggested a concept in his book called *One-Minute Manager*.[1] There have been subsequent books by others called *One-Minute Parent,* and *59-Second Employee.* How about the "One-Minute Fire Chief"?

I am not so sure it can be done. If you have read Blanchard's book, you are aware that he suggests providing praise and feedback to individuals in small doses at the time events occur. That concept has merit. But is one minute enough? Harking back to our reality issue, is sixty seconds sufficient time to capture the essence of the situation or to give an individual the

---

[1] Ken Blanchard, *The One-Minute Manager* (New York: Morrow, 1982).

feeling of praise? In actuality, I do not believe that Blanchard ever meant that the entire concept has to be crunched into that sixty-second time frame.

It does not take much of a mathematician to figure out that in a given day, there are 480 opportunities to become a one-minute manager. In any given workweek, there are 2,400 opportunities to be a one-minute manager. In a year of activity, we have the opportunity to be a one-minute manager approximately 100,000 times. The figures are mind-boggling, are they not?

Perhaps in the fire service we can accept the concept of taking slices of time to accomplish specific objectives, but I suggest that the time element needs to be structured around a concept more in keeping with the needs of the fire service. The fire service is already extremely time-oriented. How do we measure the progress of a fire? We measure it by time, of course. How do we measure the effectiveness of our response patterns in our communities? We measure it by response time. How do we measure our involvement in training programs and our various activities such as public education and fire prevention? We document the hours spent on these activities.

In the fire service, one of the most critical decision-making processes in which we engage is called "size-up." In our capacity as fire officers, most of us have had the experience of leaving the fire station knowing that we have a response time of approximately four minutes in which to arrive at the scene of an emergency. The billowing cloud of black smoke etched against the sky in daylight, or that rosy glow of loom-up at night, tells us that we had better have our act together as we arrive on the scene. Four minutes can seem like a lifetime. And in those four minutes, we try to consolidate and utilize an entire inventory of experience, knowledge, and education in order to make sound decisions under a highly stressful set of circumstances.

So, what I would like to suggest for fire chiefs is that instead of becoming one-minute managers, we become five-minute decision makers. The concept is the same as that suggested by Blanchard. As chief officers, we should focus on feedback and achievements in our subordinates and programs; however, instead of finely slicing the element of time into sixty-second portions, we can utilize the same methodology we use in controlling major emergencies. We can take five minutes to consolidate, integrate, and synthesize our knowledge and experience as we work our way through the management of a fire department on a day-to-day basis. We should be looking for technology and methodology that will assist us in using every opportunity to do this in an effective manner.

Let me be more specific about how the five minute fire chief might function. Let us take a staff meeting as an example. I once participated in a staff discussion that went something like this: As we sat around the table, it was obvious that we were going to be asked to "participate" in a decision-making process. The individual who was chairing the meeting, however, had a body language posture somewhat like a skeet shooter. You could almost hear the "click-clack" of a round being loaded into his mental shotgun as he prepared to deal with the suggestions that were going to be generated. When the problem was laid on the table and solutions were offered, true to form, the individual responded to the ideas like so many clay pigeons fluttering into the skies. Boom, boom, boom: each and every idea was abruptly shot out of the sky with a statement like "That won't work," or "We've tried that before," or "It's too costly."

A five-minute fire chief would not do that. Instead, the chief would take a sufficient amount of time to absorb each idea, allow it to incubate for a few moments, and see if it generated additional responses from other members of the staff. Instantaneous feedback is far worse than no feedback, primarily because instantaneous feedback is almost always negative. It has the tendency to inhibit the growth of ideas instead of encouraging them.

Another example of how the five-minute fire chief might function would be to reward people for performance. I recall one incident where a chief officer who attended a class on nonfinancial incentives returned to his office and immediately called his staff together, sat them down, and stated unequivocally, "You guys listen up because I am going to motivate you." Guess how much motivation actually occurred? Artificial praise and insincere pats on the back are almost worse than no compliment at all. The five-minute fire chief would take the time to ask himself, "Why did this person do what he just did?"

The five-minute fire chief reinforces the reasons that people engage in positive behavior rather than discussing the results of that behavior. It might take only sixty seconds to say, "Great job, keep it up!" But a few more minutes spent in understanding the reasons why the person engaged in the positive behavior might result in additional reinforcement, such as, "That new training program you just developed is certainly going to be of value to our recruit firefighters. What do you think we ought to do next? Why hasn't our driver training program been as successful as your new concept?" Hastily drawn praise that is too general will be lost in the conversation. On the other hand, well-developed and specific acknowledgment of successes forms the foundation for continued success in that same area.

The one-minute criticism has its disadvantages also. Before criticizing a behavior, it is best to take the time to understand exactly what is being criticized. I recall an incident in my own career involving a fire ground command that was severely criticized by my superior. I had arrived at the scene of a high-rise fire approximately ten minutes before the chief. Upon my arrival, the fire was in the flashover stage and boiling out of an eighth-floor window. A second alarm was struck. By the time the chief arrived, the second alarm companies were on the scene, the fire was totally under control, the street looked like a parking lot for used fire trucks, and not a firefighter was in sight, except for one pump operator who was providing water to the stand pipe.

The chief was very distressed to think that we had all these resources that were not being put to work. Without taking a moment to ask for an explanation, he gave a one-minute dissertation on effective use of equipment. Later, when the chief got all the information and understood my reasons for making a second alarm, he apologized for being so abrupt. The point is: One-minute feedback requires several minutes of analysis.

The five-minute fire chief does not draw and fire for effect. The "ready-fire-aim syndrome" does not result in improved performance. Instead, it results in a form of institutionalized paranoia where people are looking over their shoulders to see if the chief is going to criticize a given behavior without giving them the opportunity to explain their reasons for it.

Maybe you are already a one-minute manager. Maybe you are already a sixty-minute fire chief. In any case, the amount of time we spend directing, counseling, and providing feedback to our subordinates is essential to the overall momentum of our organizations. Whether we do it in one-minute increments or five-minute blocks of time, it is important that we apply these principles. The only thing that counts is that you have a time management system that works. With all due respect to Ken Blanchard, we do not need to be clock watchers—we need to be people movers. One person who must have his act together is the fire chief!

TOOLS OF THE TRADE

Back in the early days of my ill-spent youth, a friend and I decided to dismantle a Briggs and Stratton gasoline engine to learn about the wonders of mechanical devices. Armed with only a screwdriver and crescent wrench, we did a great job of taking it apart. Unfortunately, we could not put it together again. The reason for this is that we totally destroyed the machine by having inadequate tools to do the job. We stripped the edges of most of the nuts, and improperly slotted screwdrivers did more damage than our limited budget was able to restore. Shortly thereafter, someone told me that the real mark of professionals is not what they know, but whether they have the correct tools to do the job.

The fire service is rather tool-intensive. Big fire trucks haul large numbers of tools to the scene of emergencies. We have these tools to do special jobs at special times. In some cases, we carry tools that we never have an opportunity to use under emergency conditions.

What are the tools of the trade for the fire chief? What do they look like, and where are they stored? As I began to explore these questions, the first thing I looked at was the symbol of the fire chief's job: the five trumpets. I was struck by the fact that the symbols we have for truck officers, cross axes, and for firefighters, crossed nozzles, represent the tools of their trade. Crossed trumpets, instruments that once served as communication devices, represent the tools of our trade.

The acronym that has been used to describe the functions of top-level managers is POSDCORB, which represents planning, organizing, staffing, delegating, and so on. Those are the tasks, but where are the tools? Are there any tools we can use for planning? Are there specific tools to use for organizing? Is a budget a responsibility, or is it a tool? Perhaps it is appropriate for the chief officer to take a look at the tools of the trade—the tools he has, or should have, available to carry out the functions of being an executive in the fire service.

The tools I list here are probably not as sophisticated as some might wish. It is a possible that not all chiefs use all of these tools all the time. I would argue, however, that the tools I describe are available to almost anyone for a minimum expenditure and that almost anyone is capable of using them. The tools that I suggest are needed by an executive manager in today's setting are a Rolodex, a time management system, a goals and objectives system, and an evaluation system.

First, let us take a look at the Rolodex. If there is any one thing that we need on a daily basis, it is information. If we rely entirely upon what we already know, there is a good possibility that there will be gaps in that information base. One of the most important tools for an individual who has to cover a wide range of job knowledge is a Rolodex that reflects his experience and exposure prior to assuming the position. In his book, *How To Swim With the Sharks Without Being Eaten Alive*,[2] Harvey B. MacKay says, "If your house is on fire, make sure you get away with a Rolodex." The implication is that not only is our information network a tool, but it is one in which there is great opportunity for self-renewal. The Rolodex is very much a living tool in that it has to be added to and deleted from in order to make it relevant in day-to-day operations. Sitting alongside the telephone, fax, or

---

[2] Harvey B. MacKay, *How To Swim With the Sharks Without Being Eaten Alive* (New York, Morrow, 1988).

computer, which may be the mechanisms by which you use it, the Rolodex provides you with the possibility of finding a solution to every problem that comes across your desk.

Seldom does a fire chief face a problem that is all that unique. To paraphrase an old saying, just about everything under the sun that is going to happen already has happened. As you progress through the various ranks on your way to the position of fire chief, the names, telephone numbers, functions, and areas of expertise of all the individuals to whom you have been exposed should find a home in your Rolodex.

Our second tool is the time management system. In many columns and articles that I have written, I have emphasized the absolute necessity of having some means by which to keep yourself focused. The purpose of time management is so simple that it often goes ignored. There are only so many hours in the day. No matter how hard you try, you are not going to create any more working hours in the day. Therefore, you have to make sure that your working time remains productive and is not frittered away in incremental flights of inefficiency. If you end up working fourteen hours a day to do eight hours of work, you are going to pay a price with family and friends.

It makes no difference what type of time management system you use. Some systems are simple, such as a pocket calendar carried on the inside of a jacket, and some are sophisticated, such as computerized or electronic devices. What matters is the utility of the tool—the fact that it keeps you on track and eliminates periods of time in which you accomplish nothing.

The use of the time management tool leads to the use of goals and objectives. The concept of goals and objectives is not particularly extraordinary, nor is it all that difficult to find a system that works even for the smallest organization. The utility of this tool is that it serves as a navigational guide for all the other kinds of managerial functions. As one who likes to periodically take off on exciting, new paths that open up, I realize that goals and objectives are an important part of keeping me focused on things that should be accomplished as opposed to things that I would like to be doing.

The goals and objectives concept is similar to a firefighter's axe in that it is a tool that must be sharpened frequently. If you take an axe and continue chopping away at something, eventually the edge becomes blunt. Goals and objectives are similar in that they cannot remain the same forever. If they do, the organization becomes blunt. The utility of this tool is in always finding a new cutting edge for the organization. A fire chief needs both short-term and long-term goals and objectives, and both of these need to be reviewed on some sort of cycle to keep them relevant.

Our last tool is an evaluation system. Evaluation tools are instruments of measurement. If we were constructing a building, we would probably use a tape measure or some other form of ruler to see whether our building was measuring up to the blueprints. We do the same thing for our organization with our for evaluation instruments. These instruments can be as simple as weekly, monthly, and yearly reports that measure the activities of the organization, or they can be as sophisticated as performance evaluation systems that deal with the behaviors of specific individuals who are your subordinates. In either case, they are the means of measuring one thing against something else. In short, they are measurements of adequacy.

There are other tools that are useful for the fire chief, including such things as computers, dictation machines, and fax machines. These tools of our trade are technological advances aimed primarily at improving our ability to communicate. I do not wish to underestimate the value of these tools, but in order to communicate, we must have something to talk about. The technological tools of the fire service lie fallow unless they are driven by ideas.

This admonition applies to almost any tool. A stone chisel can be used to carve out a square piece of granite for the front surface of a building, or in the hands of an artist like Michelangelo, a stone chisel may produce a statue of David. Tools produce only according to the use we make of them.

As I recall my youthful experiment of taking apart the Briggs and Stratton motor, I remember being really interested in what was inside the engine. But not using the proper tools, I rendered the mechanical device useless. With the right tools, I may have been able to put it back together and turn it into a productive piece of equipment again. Our organizations are not too dissimilar. We may have good organizations and then dismantle them because we use our tools improperly. On the other hand, having those tools means we have the curiosity to put them to work. In the context of the fire service, we should ask ourselves from time to time whether we are using our tools to merely engage in rough carpentry or whether we are using them to produce a work of art.

GADGETS, GIZMOS, AND GOODIES: THE ELECTRONIC OFFICE

If you do not have much to do in your new job as fire chief, the rest of this chapter will not be for you. If you come to work every morning, prop your feet on the desk, open a newspaper, and read the latest news, you are probably receiving all the information you need, especially if no one ever calls asking you for input or a decision. If you have the ability to process everything that goes on in your office without benefit of interaction with anyone outside your office, then move on to another chapter. On the other hand, if you are suffering from information overload—if you are receiving input from sixteen different directions and being asked to make tough decisions frequently—it is time to take a look to see what weapons there are in the arsenal against the war on your time.

We simply cannot be everywhere at once, and many people who wish to contact us have chaotic schedules similar to our own. Therefore, we often find ourselves playing the proverbial game of telephone tag. Calling back and forth and missing one another is an extremely expensive behavior. It consumes two things simultaneously: time and money. The cost of telephone calls is exacerbated if we are talking about long-distance calls.

There are tools that we can have available to us in an electronic office that will enhance our productivity and our performance. The tools I suggest are:

1. State-of-the-art telephone capability
2. Fax machine
3. Personal computer
4. Dictation equipment
5. Paging equipment

When technology works, it is wonderful; we enjoy all sorts of benefits. But when it breaks down, we often become angry, frustrated, and vulnerable. When Bell delivered his first telephone message, "Watson, come here. I want you," it was delivered to a live human being, not to voice mail or an answering machine. Now, almost one hundred years later, when we place an important telephone call, more often than not we end up talking to an electronic device instead of a human being. The net result: telephone tag.

When we played tag, as children, the objective was to touch the other person and run away as quickly as possible. In telephone technology, leaving a message that you called and hanging up is the electronic equivalent. There are some specific techniques, however, that can minimize the impact of telephone tag. They are:

1. Eliminating "bounce-back"
2. Closing the communication
3. Making telephone appointments

State-of-the-art telephones are more than just a sender and receiver. They possess a whole array of capabilities that are important to maintaining continuity in the communications process, One such feature is call waiting. This capability to know when someone is trying to call you while you are talking to someone else is important because it gives you an option. If your conversation is important and you do not wish to stop talking, you can ignore the signal; but if you are expecting a telephone call and you are chatting socially or informally with someone, you can ask that person to wait while you take the other call.

Another feature of the state-of-the-art telephone is the concept of hands-free operation, along with the ability to conduct a teleconference. Having your hands free is not just a luxury; it allows you to operate by using your hands to conduct business at the same time that your mind is working and your mouth and ears are communicating. This is especially important if you need to find data in your desk or office in order to respond to questions or if you are conversing with more than one person at a time. The hands-free feature allows you to continue with the discussion instead of putting the other party on hold and provides a more professional atmosphere in which to continue the dialogue with people you consider to be important.

The third feature, which is incorporated into some telephone systems but must be added to others, is voice mail. Some people reject the use of voice mail outright. It is irritating to call long distance and get a tape recorder, that takes about sixty seconds of your precious time. Voice mail systems are most effective when they can be easily turned on and off so that people do not accidentally get caught in the so-called loop. The voice mail system can be used discreetly instead of generically. You can turn the feature on when you leave the office or building, and you can use it to receive incoming telephone calls when you are working on an important project and do not want to answer the telephone.

It is important that you develop telephone protocols that relate to how and when you answer telephones and that you share them with your personal and professional friends. If you have state-of-the-art telephones with features like those discussed here, you can often use them to your advantage merely by letting people know what you want them to do when they call. For example, I have often called people and told them on their voice mail that I need some information but do not need to talk to them personally. I have asked them to simply call my office and leave the answer to my question on my voice mail. Sometimes I get information from people, and yet we never actually speak to each other. All too often people fail to realize that voice mail is nothing more than another version of correspondence. If you call someone and ask them a question, it is not necessary for you to talk to them if all you want is an answer.

Eliminating bounce-back is nothing more than telling someone on the telephone that you do not want them to return the call. If you place a call to an individual and leave a message on the voice mail, the inference is that you want that person to call you back. Do not hesitate to tell the other individual that you have placed the call in an attempt to collect some information or to engage in dialogue but that it is not necessary for them to return the call. This eliminates the annoying phenomenon of ambiguous messages that waste time. How many of you have ever returned a telephone call to someone who tells you, "Yeah, I called, but I've already found the stuff I need, so I don't need to talk to you now"? Your time would not have been wasted if the originator of the communication had engaged in elimination of bounce-back.

Whenever you place a call to an individual, you should have a telephone message in mind in case you get an answering machine. I have received numerous messages on voice mail that consisted of a lot of "uhs" and "ahs" while the person tried to capture his thoughts before finally muttering a telephone number and hanging up. Having a preconceived message in your mind will help you to deal with the bounce-back phenomenon very quickly. An example of such a message is the following: "Hi, this is Chief (Blank). I placed this call to your office in order to determine what I'm supposed to do about XYZ. When you receive this message, if it is after 5:00 on Tuesday afternoon, please do not bother to return the call. By that time I will have found out the answer, or I will return the phone call to you sometime on Wednesday morning. The time of this message is approximately 1:30 in the afternoon." When you leave a message of this nature, you let the individual know that your call was important, but that timing is critical, and that if the person is unable to give you an answer within a reasonable time frame, there is no reason to call back.

Eliminating bounce-back is a professional courtesy that actually cuts both ways. It helps you to move forward in your line of thought, and it avoids a distraction on the part of the person who receives the call.

Voice mail and tape recorders also present opportunities to bring closure to communications. Frequently people place a call and, when they are unable to talk to a human being, leave a message that simply asks the person to call back. Yet the original reason for the phone call was to get something accomplished in a very specific fashion. The technique for bringing closure is similar to that of eliminating bounce-back, but instead of telling someone what not to do, this time you tell them something that you want them to do. An example of this type of message might be as follows: "Hi, this is Chief (Blank). My reason for calling is to determine what happened to the invoices that were supposed to be sent to our department by 0900 this morning. Upon receipt of this call, I would appreciate it if you would call my secretary, Mary, and advise her what time the invoices will arrive. If the invoices are unavailable, would you call us nonetheless so we can plan what to do as a backup?"

Again, when you place the call, you should have a preconceived message in your mind of what you want the person to do in case you get an electronic messaging system. I have engaged in several communications with individuals in which we exchanged very specific data without ever talking to one another. In one case, I was able to coordinate a meeting of three different individuals without ever actually talking to any one of them. If you marry this concept with another technological innovation, the fax machine, you can often give people instructions about specific information that you need them to send, and the arrival of the hard copy is confirmation that the message has worked.

The operative notion behind the idea of bringing closure to communications is "need." If you need something, you do not need to wait until the person is there. If you identify what you want, when you want it, and how you wish to have it delivered, you can often bring closure to communications in spite of the absence of the other party.

If you truly need to talk to someone and you wish to avoid telephone tag problems, you may want to try making a telephone appointment. When you call the person, be prepared to leave on the voice mail a time for them to return your telephone call. The more familiar you are with the other party, the more you can limit the number of options; however, if you do not know the person, you may wish to leave two or more times when the person could return the call so that the individual can choose one that fits with his schedule.

The key to using the telephone appointment concept is to keep the appointments. If you tell someone you will be in the office at a specific time, then be there. Always allow fifteen minutes of lag time in your own schedule. In other words, if you know you will be coming back from lunch at 1:00 P.M., you may wish to tell the person you are calling that you will be in the office after 1:15 P.M. This gives you a fifteen-minute buffer zone in case there is some kind of delay as you are returning from lunch.

These three techniques—eliminating bounce-back, bringing closure to communications, and making telephone appointments—are not foolproof. You will still have people return your calls when you have told them there is no need to call back; there will be people who will fail to do what you have asked them to do, and some people simply will not conform to the appointment process. What you need to focus on here is not their behavior, but your own. The more disciplined you are in using these techniques, the more likely it is that you will control the communications process. Engaging in these techniques will definitely lower your anxiety level because you will be taking progressive steps that do not rely upon the other party returning the call before you can move on to other elements of your day-to-day problems.

Individuals who develop these skills can convert telephone technology into just another mode of communications. If you are successful using these techniques, you might be able to convert telephone tag into a spectator sport.

This leads us to the next technological tool on the list, and probably the most ubiquitous new device in the field of electronics, the fax machine. There are pros and cons about owning a fax machine or having one in your office. I will not debate the negatives because those people who do not want one simply do not understand the primary advantage of the fax machine. It is a time saver. You can actually use it to accomplish two things. First, you can call people, tell them you need information, and instruct them to fax it to you directly; you do not even need to talk to them. Second, you can send a fax to people and tell them you need to make an appointment to talk with them at a specific time so they can make themselves available by telephone instead of wasting time talking to their voice mail.

Both of these applications of the fax machines can be used to enhance your ability to control the communications process. If you sit in your office with your fax plugged in, just waiting for it to come alive and transmit some mystical message from the outer world, you may find yourself sadly disappointed. People will not "fax you" unless they know that you have the machinery. Unless you establish communications with other people by using the fax, they may fall back upon telephone tag to try to work things out with you.

I have engaged in a series of communications in which I called someone and left a message on the voice mail telling him that I needed something very specific and requesting that he fax it to me as quickly as possible. That person faxed me the information along with some annotations at the bottom of the paper advising me of someone else who had information. I then faxed the original correspondence to the second party, requesting supplemental information, and received an answer from that person in less than thirty minutes.

Another technological tool is the personal computer. To many chief officers, the idea of having a computer on their desk is anathema. It takes on the connotation of performing secretarial-type functions. But that is not the purpose for which I suggest that it be used. Instead, the personal computer on an executive's desk should be the executive's information retrieval tool.

A personal computer is an excellent device with which to store database information on telephone numbers, names, addresses, contacts, and so on. My personal preference is to use a software program that contains the names, addresses, and phone numbers of people who are involved in my information exchange network. I frequently go to the "find" mode on the computer as I am conversing with someone on the telephone. For example, if someone mentions Chief XYZ, I can type in Chief XYZ's name and instantly pull his name, address, telephone number, and other information up on the screen. This often gives me an additional clue to how that information might be useful.

Database management is somewhat complicated, and I am not suggesting that the fire chief get heavily involved in using the computer to operate his entire life. However, the personal computer is useful for creating master calendars and keeping track of events. It allows you, with a few keystrokes, to collect information on a day-to-day basis that can be summarized later to reflect your productivity and effectiveness.

Closely aligned with the personal computer is the concept of handheld executive organizers. The type that I use is about the size of a handheld calculator, and it contains almost all of the functions I need to keep organized on a day-to-day basis. For example, it contains a calendar program, a memo writing program, and a to-do list. I have almost two thousand telephone numbers in that pocket organizer. There are interfaces that allow you to download this information directly into the software of your personal computer, which means that the information you collect in the field is instantly added to your files in the office and at home. You can also download the information on your personal computer to the handheld organizer.

E-mail may be viewed as either a curse or a blessing. The purpose of e-mail is to allow the transfer of information between individuals without personal contact. It is not supposed to be used as a "gotcha" device or as an excuse maker. E-mail should consist of a brief exchange that either asks a question or provides an answer. It should not be used as a way of avoiding personal contact with people. E-mail overload frequently occurs when everybody copies everybody else on their e-mails, a form of defensiveness that I consider to be offensive. When you use e-mail, consider it like a conversation, not a trail of evidence. Although I have heard a lot of criticism about e-mail, I have found it to be a very useful tool for coordination and closure.

The next topic for consideration is the use of dictation equipment to improve your use of down time. One of the most difficult things in the world to do is to convince fire chiefs that they need to learn to use dictation devices. For some reason, fire chiefs are almost intimidated by the idea of talking to equipment instead of human beings. Yet, dictation devices are the best time-saving machines you can learn to master.

The only way to become accustomed to using dictation machines is to pick one up and start talking into it. You might want to start with an outline. I frequently make notes to myself and tape them to the dashboard of my automobile so that I can dictate letters, notes, and articles, while I am traveling to meetings. I have often used my dictation machine to give instructions to my secretary when I knew that we would not have face-to-face contact.

Sometimes I have left reports on my secretary's desk with a tape giving an overview of what I wish to have done with them. The point is, a dictation machine is a communication device. It is a time-saving device that allows your voice to communicate to someone who can transcribe, interpret, or otherwise follow through when you are not there to do something yourself.

The pager is another technological device around which there is a considerable amount of controversy. The idea of carrying a pager has gone through the various stages of being unique, followed by being somewhat of a status symbol, followed by being a real pain in the neck. If you are interested in vital communications, however, you may wish to consider the pager as a tool to keep you in pace with reality. For example, I have left my office knowing that I would be receiving an important phone call. I have given instructions to my staff to put the person on hold and to call me on the pager. Then, utilizing either a mobile telephone or the closest manual telephone, I would have the phone call patched directly through to me, and the communication link was forged.

The thing you must remember about paging systems is that they are not there to serve as an annoyance. You should limit the number of people who know how to page you, and you should limit the circumstances under which they are allowed to page you. I have told my staff that they are not to page me for routine details that require only a yes or no answer. I make it my personal practice to advise my secretarial staff when I am actually wearing the pager and when I am not.

So, where will all of these electronic gadgets, gizmos, and goodies take us? Is the fire chief of the future going to be a bionic cyborg of electronic communications or an individual who is in control of his own destiny? Each of the electronic devices described here has a cost factor and not everyone can afford all of them simultaneously; you may need to pick and choose. But these devices, although intimidating at times, are tools that will allow you to control the communications process on your own terms. Instead of being overwhelmed, I often feel greatly relieved when I use these tools, and I have more spare time to choose between alternatives instead of spending all my time trying to fill in the blanks because of missed communications.

If you are a highly involved individual who is making a lot of things happen in your organization, you may have found yourself thinking, during this discussion, "Yeah, there's a couple of times when that sure could have helped me out." This is the kind of thinking that is important for the fire chief. We are at war with a host of adversaries, including fire, misuse of our time, and the stresses and strains of keeping the fire service moving forward. Not unlike Captain Kirk standing on the bridge of the Enterprise, we need to be able to communicate with the greatest of ease and the highest level of confidence. The electronic office is not a luxury; it is an arsenal.

TAKING MINUTES AND WASTING HOURS

I once read a definition of a meeting that went something like this: A meeting is an event in which you spend many hours of wasted time so you can produce "minutes" later on. Many individuals complain about the fact that there are too many meetings and that some of what is discussed at them is absolutely irrelevant. Yet, we all recognize that we have legitimate reasons for getting together with people and that it is important for us to work together in a teamlike atmosphere. So the question is not whether we have meetings, but how to make these meetings more effective and relevant to our daily activities.

It might help to take some time to explore the purpose of meetings. There are many ways to classify meetings. They can be classified as being essentially one-way or two-way interchanges. Sometimes we hold a meeting to merely pass out information so that everyone is kept informed, and sometimes we hold a meeting so that there can be an exchange of information between individuals who are attempting to work on similar problems. Meetings can also be classified as regularly scheduled or intermittent. The more regularly scheduled a meeting is, the more there has to be something meaningful on the agenda to be discussed. The intermittent or unscheduled meeting is generally successful only when there is just a single item to be addressed.

Let us review the parameters of what constitutes a good meeting. The first parameter is that there must be a good reason to get together. The most successful meetings focus on specific topics. Ask yourself this simple question, "Why are we getting together today?" If you can answer it, you are halfway home to having a successful meeting.

The second parameter is expressed in the question, "What do we intend to do after this meeting is over?" If the answer is that you are going to make a decision, and take a specific course of action, then the meeting stands a high degree of possibility of being successful. If the purpose of the meeting is to do nothing more than to defer a decision, why have it?

Another good question to ask is, "Who needs to be at this meeting?" All too often meetings are held to bring together large numbers of people, most of whom have nothing to contribute. One of the most courteous things you can do for your staff is to make sure that if they come to a meeting, they are included in it. One of the most discourteous things you can do to your staff is invite them to a meeting in which they have absolutely no concern, and then make them sit there and listen to an exchange between one or two other people. The idea that meetings should be conducted between people with relevant relationships is one that many individuals totally overlook.

Another question to ask regarding meetings is, "Who will be responsible to take action?" The worst sin in holding a meeting is to fail to make a decision. The second worst sin is failing to identify the person who will implement a decision that is made. Any time you finish a meeting by saying that a decision has been made but not saying who must take action, there is the possibility of a slight hitch in bringing closure.

Then there are the actual logistics of meetings. There has been a lot of ink spilled over the topic of how a meeting room should be arranged—whether it should be horseshoe-shaped, whether the tables should be put in a circle, and so on. Frankly, I think that arrangement is overemphasized and entirely too mechanical. It does not make any difference how the meeting is laid out as long as it is functional. Any time you are holding a meeting to deal with specifics, all the information needs to be right there at the time you introduce the topic—charts, graphs, illustrations, bulletins, and maps are of absolutely no use to the decision-making process if they are stuck in a drawer or out of reach.

Similarly, it is not always necessary to be overly concerned about the creature comforts as long as the meeting is of reasonable duration. Long meetings often require all sorts of accoutrements to make them comfortable, but if you focus the intent of the meeting on a very specific item, the room furnishings take a backseat to the decision-making process. One of the most functional meetings is the stand-up meeting. This is where you call a group of people together in a small area and say, "We have a problem. I need your input. What do we need to do here to fix this?" Then you move on.

Granted, it is not always cut and dried. There are certain meetings that need to be held on a regular basis if for no other reason than to calendar and coordinate the activities of the members of the organization. My personal preference is to hold what I call a calendar meeting every Monday in order to let everyone know what I will be doing that week and determine what everyone else will be doing during the same time frame to see if we can avoid conflicts, duplications, overlaps, and omissions.

Probably one of the most abused forms of meetings is the "staff meeting." Everyone is brought together for the purpose of having the boss tell everyone how the cow ate the cabbage. Staff meetings are almost always shaped like a pyramid. Large masses of people sit in the audience while a small number of people actually make contributions and discuss things. The greatest danger of staff meetings is that they can become so monolithic and boring that the individuals at the bottom simply get nothing out of them. The greatest strength of a staff meeting is revealed when you turn the pyramid upside down and ask the individuals who are brought to the meeting to contribute their thoughts, ideas, and concerns about the activities in the organization.

There are people who come from the school of thought that the fewer meetings you hold, the better off you are. There are also people who believe that you have to meet on practically everything because they have read too many books on participative management. The answer is probably found somewhere in between these two extremes. You do not need to hold meetings to have communication. You don't need to hold meetings for ceremonial purposes. The more that a chief officer or supervisor uses meetings as a point of contact with an emphasis on action, the more likely it is that the meetings will be meaningful and people will actually look forward to participating in them.

One other thing we must be very careful of in regard to meetings is to make sure that meetings not be used as a form of punishment. Frequently meetings are held in order to air dirty laundry or to hold people up for ridicule. Not only is this counterproductive and a waste of time, but it also has an adverse impact when people return to the workforce. Any time a person is brought in front of a group of people at a meeting and given some form of verbal discipline for failing to achieve something, the other individuals return to their workplace asking themselves, "Will it be me next time?"

Here is one last suggestion about meetings: A person who calls a meeting must accept responsibility for that meeting. If you are in a superior position and request your subordinates to attend a meeting, it is up to you to make sure that they take something away from it. If they have requested your time and asked you to attend one of their meetings, it is their responsibility to make sure that you get something out of it. If everyone who sits around the table or stands up to engage in a dialogue under the guise of a meeting remembers that there is a mutual responsibility to give and take or there is no reason to hold the meeting, then things cannot help but improve.

There is a concept in decision making which says that every decision can be anywhere from totally democratic all the way to totally autocratic. You must remember that when you conduct meetings in which you are asking for input, it is a consensus-gathering process. Individuals who participate in meetings must realize that when their input is requested, they are not being asked to vote on something. Often the responsibility to take action is in the hands of the individual who is conducting the meeting. That is not being autocratic—that is being responsible.

I am sure that many organizations will still have frivolous meetings, and I have no doubt in my mind that many people will still sit in meetings drawing doodles on paper and wishing they were somewhere else. If those who are responsible for conducting meetings, however, pay a little more attention to the ground rules discussed here, the opportunity for that will probably diminish, and we will begin to see an increase in the effective use of our own time.

SELECTED READER ACTIVITIES

1. Evaluate your own point of view regarding technology. Of the four types of orientation, which one best describes you?
2. Review your management and leadership style. Evaluate your approach to both decision making processes and feedback to your personnel.
3. Examine your own use of time management techniques and planning processes. Are there areas in which you could use improvement?
4. Create and maintain your own personal Rolodex or telephone directory.
5. Review the manner and method with which you conduct meetings. Are the meetings valuable? Are commitments made and kept? What is the value of the meetings, and what are the consequences of not having them?

STAFF MEETING TODAY 11:30...

YOUR STAFF

DAVE HILBERT 9?

CHAPTER 12

High-Performance Teams:

Getting a Hold on the Process

Based on the last chapter, once you have taken control of your own use of time and resources, you must turn your attention to taking control of the organizational processes. The emphasis in this chapter is on group dynamics and team management, using a range of behaviors from the warm and fuzzy participative style all the way through outright dictatorship.

> *An army of sheep led by a lion would defeat an army of lions led by a sheep.*
>
> —*Arab proverb*

CHOOSING SIDES: DEVELOPING THE TEAM

Early in life we learn a valuable lesson about teamwork. Unfortunately, many of us lose that perspective as we grow older. I am referring to the act that takes place on many a playground or school ground when it comes to choosing a team. There is a process utilized even by young children that has a lot of merit. If you witness what happens with small children in choosing teams, you will notice an interesting phenomenon. In the first place, the team captains are usually chosen because they are leaders.

In my neighborhood, we used to flip a coin or use hand over hand on a bat for the first pick. Everyone else was chosen in sequence. The two leaders usually picked the two best persons for their team, and then they alternated picking the remaining participants until the teams achieved some type of equitable distribution. If the selection was based on competency alone, it was not uncommon for the two teams to be relatively equally matched. When, and this happened too frequently, the choice was made on personality alone, it was not uncommon for one team to be overwhelming in its ability to compete.

How does this relate to the modern world of fire protection? Developing your team is a very important aspect of managing a community's fire problem. If your team fails to have a specific talent or ability that is required, it is possible that you will lose the game. If a team is well balanced, it can become quite competitive.

In the first place, there are leaders on both sides of any fire protection policy discussion. We consider ourselves, as fire chiefs to be the leader of the advocacy role. On the other hand, we may have opponents who have very capable leaders. Without pointing the finger at any specific industry or occupation, it is clear that in the field of fire protection, fire chiefs have their detractors. The more effective these detractors are in carrying out their agenda, the more difficulties we face in achieving ours.

Secondarily, in the fire profession, we have a strong need for a range of skills and abilities to make the team work. Throughout the history of most fire departments, the personnel who have staffed the team tended to come from a very narrow spectrum of individuals, mostly homegrown. With the continued evolution in the complexity of fire protection today, your team should not come from a narrow focus of individuals if you can avoid it.

We would not want a baseball team that consisted of all pitchers and catchers; the team needs to have utility infielders and a coach. The same thing applies to our fire departments. The fire chief cannot be everything to everyone in the context of the organization. Although there is some general knowledge required for the job, such as overall familiarity with management and leadership, the Chief needs to be surrounded by people who know more than he does about specific topics. There are some distinct differences in the specialization of people who serve as training officers, operations chiefs, and fire prevention personnel. Even within a fire prevention bureau, plan checkers can sometimes be more specialized than field inspectors. Field inspectors may be more specialized than public education officers.

Many people reject the idea of specialization. The vast majority of fire chiefs in this country probably have fire staffs that are very small. The case I am making here, however, does not address whether a person has specialists on staff but, rather, whether a person needs to develop specialties among subordinates instead of shouldering all the responsibility himself. In a smaller agency with only one or two staff officers, a fire chief may have the same skills and abilities as his subordinates, which sometimes results in the chief not properly delegating duties. The other staff members become observers instead of helpers. As a department grows in size, the diversification of skills and the development of a "team approach" can do nothing but strengthen the authority that has jurisdictional ability to carry out its mission.

There are many things that need to be considered by fire departments in developing this teamwork approach. One of the best ways to keep this in proper perspective is to look at the fire department team as having different positions to be played. By way of the sports analogy, the fire chief may actually be a person who has to serve as a playing coach. The nature of most teamwork is to try to put together a game plan that results in winning. In the context of fire departments, there are not a lot of individual winners and losers on a day-to-day basis. There are only organizational winners and losers based on successes and failures of the entire delivery system. Not all of the successes or failures of a fire chief are measured at the time of an emergency.

The more that the chief of a fire department concentrates on the development of a spectrum of teamwork capabilities, the more likely it is that the team will produce outstanding results on all activities. Job descriptions that are almost cookie-cutter impressions of one another with only slight variances in the level of responsibility breed mediocrity. So, one technique a chief can use is to make sure to review all job descriptions of subordinate staff. When we were back on the playground at the age of nine or ten, we always tried to pick the best team members that we could so we would win. Nothing felt better than to be one of the first people selected; conversely, nothing felt worse than to be the last person picked.

The more we focus upon competency as part of our teamwork, the more we will find people striving for a level of excellence in order to be part of that team. We might be surprised when we discover to what extent we already possess the our human resources needed to complete the team. Whether a person is an entry-level firefighter or a seasoned fire officer, most of our staffs come with a desire to serve the community. There are many potential team members; you just need to find them and put them to work.

TEAM: THE ACRONYM

The word *team* means a lot of different things to different people. It is used most often in the context of team building, teamwork, or team efforts. The implication is that if you are on the team, you are part of the decision-making process, but if you are not on the team, your efforts are either counterproductive or at least minimized.

Dilbert the cartoon character, has lampooned the concept of teamwork a great deal. His character infers that what we call teamwork in management is really just another term for managerial intimidation. After all, most of the people who talk about team building and teamwork seem to be at the top of the pecking order. Those at the bottom seldom see it as a concern. More often than not, those at the bottom are more interested in knowing what is going on, what are their chances of survival, and whether they might be promoted.

Teams are a great concept, but what do they really mean to us in the fire service? Is a fire company a team? Is a battalion a team? Is a platoon a team? Is the department a team? Is the department part of the city's team? These are good questions, but they are hard ones to answer. Most organizations struggle daily to find out whether they are succeeding. Furthermore, being a team at one level does not ensure that people are collectively a team at the next level.

A friend of mine once gave me a card that used TEAM as an acronym for Together Everyone Accomplishes More. The phrase is catchy, but if it is true, then why are people not more willing to work together? This acronym contains a fatal flaw. Together everyone can accomplish more, but we do not reward everyone equally for achieving something. To the contrary, we often set up systems that reward only those at the top and ignore the contribution of those further down. A classic example is that inconsistency of the so-called pay-for-performance system—the rewards programs that can be given to only a percentage of the workforce. Together everyone produces something, but only a few get compensated.

There is another glaring error in the idea that togetherness produces results. Too often, togetherness produces "group-think." That is the phenomenon that occurs when everyone is so compatible that they are not aware of gaps in their own thinking. Teams that consist of people who all think alike can suffer from a form of creative nearsightedness that leads to monumental failure.

Let us take a step back from the organizational development rhetoric about teams and look at the concept in a more fundamental manner—from the perspective of personal participation. We are all familiar with the idea of teams in a sporting sense, and there are some good lessons for us in that use of the concept. Factors that apply to sports teams that bear some consideration by those of us in the governmental arena are as follows:

1. Games are played according to the rules.
2. There is a field on which the games are played; hence, you cannot go out of bounds.
3. There are referees and umpires to check on numbers 1 and 2 above.
4. Teams are either on offense or defense, but never both at the same time.
5. Everyone on a team is equally important because a specific play can require any team member's specific knowledge, skill, or ability to score or stop the other team from scoring.
6. Teams always keep score to determine who wins and loses; in every game, there is a winner and a loser.
7. When the team is a winner, everyone wins individually; when the team is a loser, everyone loses individually too.

Do all of these factors apply in the context of government service? My answer is yes, no, and maybe, depending on a number of things. But let us start with this basic statement: you cannot have a team unless everyone accepts the idea expressed in number 7 of the list. Everyone wins, or no one does; everyone loses if anyone does. Take a moment to think about that statement. Daily, weekly, monthly, and annually we have situations in the fire department where winning and losing is the source of conflict. You can probably recall events from the recent past when members of your own team were in conflict with one another or with you. Sometimes there are multiple conflicts occurring within a team as well as internecine warfare between stations, shifts, divisions, and so on.

How can we succeed in winning our overall game of fire protection if such conflict exists? Can we ever hope to win if we cannot eliminate the conflict? Here again is that interesting answer of yes, no, and maybe. Yes, we can be more successful as a team if we look at our differences of opinion as assets, not liabilities. No, we will never show improvement if conflict is allowed to be personalized. Maybe we can show improvement as we start to think of a new way of looking at a team.

I would like to offer TEAM as an acronym with an alternative interpretation: Total Empathy Among Members. What does *empathy* mean? It sounds like sympathy, which is essentially your feeling sorry for someone. Empathy means that you understand the other person's point of view. In essence, teams made up of people who think that efficiency can be developed in spite of differences of opinion and who understand one another's point of view can often collectively resolve issues and problems faster than any other kind of team.

People with empathy attack problems instead of each other. What a simple concept, but what a difficult task. It is so difficult that some organizations are totally incapable of achieving that level of relationship. Some conflicts result in the destruction of individuals and organizations. This destruction is equivalent to losing the game. There are no individual winners when a team fails; the other side wins, sometimes by default.

Let us go back to the list of factors that help sports teams to succeed. By rewording the statements, that same set of guidelines can help us in the fire service:

1. Our team has a set of rules by which we play the game.
2. The field of play is internal to our department, division, company, and so on.
3. We actively seek outside input from our stakeholders before we change either the rules or the playing field.
4. We have two sets of strategies—proactive (offense) and reactive (defense)—and we keep them separate.
5. We value every member of the department for that person's potential ability to contribute to this organization's success.
6. We tell everyone what we plan on doing, and we measure what we did; we keep score.
7. We attack problems, not each other—we celebrate our success at all levels, and we mourn our failure at all levels.

These are conceptual ideas that need to be refined in terms of individual behavior before the concept can become reality. A good place to start describing the behavior to the team is to ask the members to listen to each other before deciding who is right and who is wrong in areas of disagreement. With the paramilitary structure of the fire service, we often equate rank with knowledge, wisdom, experience, power, authority, and good judgment. This assumption can fool us sometimes. Newer members of the team do have to learn about things before being allowed to turn them upside down by youthful exuberances and incipient enthusiasm for untried ideas, but total empathy among members means learning to separate facts from opinions. It also means respecting that facts can be discovered during an exchange of differences. Empathy comes about when people give other people the benefit of the doubt, before doubting the benefits of what they have to offer.

Teams begin when individuals appreciate one another's strengths, but they become overwhelmingly successful when each person begins to protect his teammates from their own weaknesses. Teams come together when they are doing well, but stay together when they are doing poorly. Teams focus upon the team's reward instead of upon the individuals' performances. Teams play best in the public eye, and they criticize themselves behind closed doors. Teams are stronger than any one individual, including the coach, and are as weak as anyone who is allowed to function without help from the others.

In my career, I have been privileged to be part of several team efforts that have been successful, and I have also been on teams that have failed. No doubt many of you have had similar experiences. What seems to be changing for many of us today are the stakes of that process. We keep score in different ways, but one that is important to me is the role of preserving our profession and protecting the lives and property in our communities in a way that indicates that we are the champions of life safety.

The Super Bowl in our business does not always occur on a certain date each year. It can happen at budget time or at the time of the most catastrophic event of the century. Or it can occur when we least expect it—when we solve a major problem or remove a potential controversy from our department and do so in a manner that earns us the respect and admiration of even our staunchest adversaries. Then, instead of responding to a five-alarm blaze, we can all respond with high fives to each other because we have truly functioned as a team.

FILTERS, FUNNELS, AND CLOGGED DRAINS

The primary purpose of a funnel is to be able to pour liquids, or even dry powders, from a larger container into a smaller container without spilling them all over the floor. And if you have an impurity in the commodity that is being poured from the larger container to the smaller one, you can use a filter to eliminate the impurity or contamination in the product. Sometimes the material that is being screened through the filter reaches a point where it clogs the screen, and nothing can get into the subsequent container. You have to stop and clean the filter.

This is comparable to the process of communications in many organizations. Every level of rank in the fire service represents a funnel, and the decision to communicate or not communicate represents the filter. If the filter becomes so clogged that it fails to transmit the intended communication, the process results in failed communications.

A candidate for the position of chief once told me that he perceived his role as the commanding officer of the organization to be both the funnel and the filter. Those were not his exact words, but that was what he was saying. He indicated that whenever the chief officer gave him information, he felt that it was his job to make sure that the information was properly packaged, channeled, and utilized by his subordinates. He recognized that sometimes he was made aware of things that did not need to be communicated to other individuals in the organization because the information simply was not relevant or because it was so tentative that it would cause more problems than it would solve if it were distributed.

When you are the chief of the department, you need to feed the funnel, and sometimes you have to clean the filters. You need some command officers working underneath you who you can provide a large amount of information and who can channel it correctly down through the hierarchy to the appropriate level in the organization to get things done. These individuals who are responding to you need to know how to discriminate between what is appropriate to channel and what is not. If these capacities are not in place, your organization will suffer.

Anytime a command officer working for you does not have a filter mechanism in place, you have a problem. Many of the discussions that you will have with your immediate subordinates involve details that are extremely sketchy, tentative, or even confidential. How they treat that information either helps or hurts the organization in dealing with change. Sometimes we are researching a project and have incomplete information; sometimes we are flying trial balloons to get responses and are interested only in immediate feedback, and sometimes we are merely asking for input before we get ready to make a decision. Therefore, our immediate subordinates have the opportunity to do one of two things with this kind of information: they can engage in a meaningful dialogue with us so that we can make better informed decisions for the department, or they can take the information and translate it, interpret it, and transmit it to other members of the organization with a specific spin on it that can cause us a lot a grief.

How do I know this? Because I have seen it happen. The relationship between a top-level chief officer and his immediate subordinates is an extremely critical one for the establishment of communications in an organization. This relationship tends to affect the entire morale, if not the overall performance, of the organization. When the relationship between chief and staff is strong and the use of the communications process has a great deal of internal integrity, the organization tends to be strong. The opposite is also true. When the funneling and filtering system in the organization is clogged or the filters are not in place, there is a possibility that the organization will suffer.

As chief officers we need to recognize that this filter and funnel process applies to us as well as to our subordinates. When it comes to dealing with our immediate superiors, we have to funnel and filter information that we receive from them in a way that is not contradictory to the working relationship with the authority having jurisdiction. The most effective communicators have a lot of credibility because they practice what they preach. They listen carefully to the directions of their superiors and utilize the filtering system and the conversational process to ensure that they are communicating effectively to their subordinates. On the other hand, there are individuals who seize every opportunity to be inconsistent, that is, they demand that their personnel remain intensely loyal to them and perform in a certain way, yet they are disloyal to their immediate superiors.

The communications process is complex at best. There are many elements, such as choice of vocabulary and body language, that affect communications. But the focus here is on one specific phenomenon: the responsibility of command level officers to serve as a conduit of information and to have the insight to know what to communicate. The following considerations in this area might help us as fire chiefs in relationship to our immediate subordinates.

First, openness and candor do not necessarily mean telling everything to everybody all of the time. It is appropriate that we maintain confidentiality in our relationships and that we are open about the fact that confidentiality is part of the process. I have held many conversations with individuals in which I told them that the information being discussed was to remain between the two of us until such time as I indicated it was appropriate for open discussion. Most of the time individuals have complied with this request, but sometimes they did not. Confidentiality is a form of trust. It does not mean that you are being secretive, it merely means that the information being utilized is not appropriate for distribution throughout the organization where it is likely to feed rumors and to cause anxiety.

Second, it should be permissible for people to ask whether or not something is appropriate without looking foolish. If you are engaged in a dialogue and someone is concerned that the information that he is being given may have some implications elsewhere in the organization, he should feel perfectly free to speak up. We should not be the least bit offended when that happens.

Third, you should reserve the exercise of confidentiality for items that are specific and intrinsic to the security of the department. Discretion is extremely important because if everything becomes confidential, people will begin to believe that they are being held out of the information loop.

If you are open and candid with members of your organization, there is a high degree of possibility that you will get feedback and compliance from them. The funneling system goes from top to bottom, as large amounts of information that come into the top are filtered down and then utilized by other members of the organization, but there is also a funnel from the floor that comes up to us. We have to depend upon our command officers to filter that information as well so that only important issues come up through the process for decision making.

One might visualize this communication process as two radio channels, one of which is sending, and the other of which is receiving. We can use the funnel to provide a lot of information to our people about our expectations, but we must use our filters to make sure that we talk to them about the right things at the right time. We can also use the funnel to collect large amounts of input from our people so that we can make better informed decisions at the top, but we must rely on our command officers to filter the information to make sure that we are using our time intelligently. Failure to use this process properly can result in a "clogged drain" in our organization; and we all know what happens then—things back up. Unfortunately, there are some things that will always flow uphill, however, and bad news is one of them.

LABOR-MANAGEMENT RELATIONSHIPS: A NECESSARY EVIL?

We might as well start this section with a controversial statement. It appears that most labor problems in fire departments are caused by *bad management practices;* conversely, many management conflicts are created by *bad labor relationships.* Now that I have offended a large percentage of people, let us explore the problems addressed by this statement.

The statement addresses a major issue in many fire departments: the type of relationship between the working force in an organization and the management and leadership in that organization. I am not referring to labor union relationships because a labor-management relationship exists whether or not there is an organized labor force. One might think that the existence of organized representation would result in better relationships, but in many cases it does not. Nor does it result in more successes at the bargaining table. In addition, there are many cases in which strong labor leaders have taken control of some management decisions in organizations. But this section is not about the legal aspects of labor relations or the reasons why individuals belong to labor organizations. Rather, it focuses on the one word that seems to get dropped from the conversation when we are talking about labor and management: *relationship.* For it is the ability of labor and management to communicate, coordinate, and, in some cases, share different agendas that often results in the strength or weakness of an organization as a whole.

In my career, I have had the opportunity to sit on both sides of the table on this issue. Early on, I worked on the creation of a bargaining unit and served in a leadership role in a labor organization. Then, as I engaged in other career development and professional opportunities, I found myself in a management role sitting on the opposite side of the table with some of the same individuals with whom I grew up and with whom I shared many common experiences. My experiences are not particularly unique among fire chiefs; many of them have commented that some of the best training for leadership in a management role has come out of the experience of working in a leadership capacity of a labor organization.

Then why is the term *labor-management* often a precursor for terms expressing conflict and advocacy situations? Is there enough room in organizations to have generals and privates, chiefs and firefighters, and yet share a common goal and desire to see the organization move forward? I certainly hope so, because it is the establishment of relationships between the backbone of an organization and the nerve center of an organization that allows it to become an influence in the profession.

The fire service is not all that different from a military organization or a professional athletic team that is attempting to win a battle. The subtle difference is that labor and management in the fire service often consist of individuals with almost identical backgrounds and yet with different perspectives on where the organization is going. Over the last couple of years, there has been a major change in the focus of fire department administration that may help to resolve this relationship issue. More and more fire departments are creating mission statements and identifying goals and objectives, Fire departments are beginning to look at the overall organization as a total system—an organism that is designed to accomplish specific tasks, instead of two separate groups that only get together to discuss the fringe benefits and compensation of the organization.

Since this book is written primarily for fire chiefs or for those aspiring to become fire chiefs, I would like to discuss the improvement of labor-management relationships in the fire service from that perspective. There are no magic formulas or quick fixes, but a focus on three aspects of the fire chief's role can result in improved labor-management relationships in an organization. These are aspects that are not addressed in courses on tactics and strategy, that cannot be found within the pages of a fire prevention and code enforcement manual, and that are often omitted from the job description of a chief officer. These aspects are principles, honesty, and credibility. A fire chief must adhere to all of them.

Let us explore adherence to principles first. We must admit that although labor and management in the fire service share many common factors such as background, training, education, and even personal experience, there are definitely different agendas for labor and management personnel in an organization. Principles are the things for thich we stand. To use an old cliché on this topic, "Unless you stand for something, you'll fall for anything." In establishing labor and management relationships, it is important that these principles be established in some respect at the very outset. Even though individuals may differ, as reasonable people often do, on issues such as adequate compensation, rules of fair practice, and other sets of rules, often there is concurrence between individuals of different missions on things such as the need for a set of "game rules" by which to play the game.

In the most hostile of adversary relationships, war, humans have managed to establish the Geneva Rules of Warfare. Even in this situation of conflicts that is extreme enough to cost lives, there is a set of principles upon which civilized nations can agree. Yet, in the operation of labor relations in an organization, often there is no set of ground rules to control the behavior, response, reaction, and open conflict that occur as a result of labor-management relationships.

It is most important in the formation of a relationship between labor and management that the ground rules be spelled out. These have to do more with areas of common ground than they do with areas of differences of opinion. The principles of good labor relationships are really the principles of good negotiation. In many cases, labor organizations have conducted extensive training and workshops for their leadership in order to provide them with the tools they need to engage in good negotiations. Many individuals, however, assume the role of fire chief without having an adequate amount of exposure to the negotiations process or the tools of the trade used by professionals in the field of conflict resolution. These individuals too often interpret things personally instead of professionally.

The issue of honesty may appear on its surface to be a given. Unfortunately, in establishing labor-management relationships, there is often a dimension to honesty that is often missing, and that is candor. It is all right to disagree on issues if people will simply state their positions and articulate their differences at the outset instead of saying one thing in face-to-face communication and another when discussing the issue with third parties. Honesty is not something that is achieved easily in the negotiation process because negotiations often have an inherent dimension of withholding information in hopes that the other side will agree to something. But honesty does have a place in this environment.

The fire chief's role in making sure that honesty is maintained is simply one of keeping the hand on the rudder of the organization. Granted, many labor-management relationship scenarios result in bluffing, posturing, name calling, and a wide variety of other types of psychological warfare. But from the fire chief's point of view, the most stable position from which to function is one of absolute honesty about what you are expecting to obtain in any given labor-management relationship. I have seen several fire chiefs derailed in this process because they said one thing to their labor group and quite another to city management. As a direct result of leaks in the system, their honesty was impugned, and they were unable to achieve their objectives.

This leads to the last of the three characteristics, which is credibility. This dimension is difficult to describe. It is almost like art: I do not know what it is, but I recognize it when I see it. Credibility as a dimension in labor-management relationships is essentially a measurement of the degree to which a person is true to himself and to his word. Credibility involves things such as never making promises that you cannot fulfill or that you never intend to fulfill.

In society, we recognize credibility to a degree by the way we accept various individuals in the advertising field. For example, Bill Cosby is an actor and comedian with a doctorate degree in Education. He is used as a spokesperson for a lot of products. His reputation as a person is translated from the screen to the buying practices of many individuals because he is "credible." People buy products that he recommends and accept his advice on taking actions that are different than the ones they would choose for themselves because they think he has insight that they do not possess.

In a fire chief's role, credibility is not something that can be achieved in an instant. It is something that accumulates in an organization over a period of time as a person works out his labor-management relationships and establishes the fact that he is who he claims to be and that he is consistent in his application of labor-management practices.

The actual manner in which labor-management relationships are conducted in individual organizations is all over the map. In some cases, the relationship has been established in a hostile environment. In other cases, it has been established in a relatively benign vacuum. In almost all cases, labor-management relationships either get better or get worse. They seldom remain at a status quo for very long. The various conditions that exist in a community such as the economics, growth, increase in challenges and opportunities facing the fire department, change in demographics, and a wide variety of other influences ultimately affect labor-management relationships.

But more importantly, these relationships are not unlike the proverbial rose garden. Although the gardener may plant beautiful bushes that will produce blossoms under certain conditions, a rose garden has to be weeded and fertilized almost constantly to remain a vision of beauty. And so it is with labor-management relationships. There is no such thing as establishing the relationship and allowing it to sustain itself. Maintenance of the relationship requires give and take. It is a communications process in which labor personnel must feel free to express themselves and their issues without fear of reprisal, and management personnel must be free to state their vision of the future and their desire to direct the organization without fear of recalcitrant and unwilling support.

In the best of all possible worlds, labor-management relationships result in a type of homeostasis that allows an organization to move forward evenly over an extended period of time. In the worst possible scenario, labor-management relationships become a battleground of winning and losing in which terrain is constantly lost by one side or the other. In military campaigns, it is seldom that a general will lose a battle while the personnel will win the war. And so it is with the fire service. It is virtually impossible for the fire chief to take a position in a labor-management relationship that is so contrary to his organization that the chief can win and the organization cannot lose.

Some of you may regret the fact that the fire chief ever has to be involved in labor-management relationships; and in some cases, this responsibility has been removed from the fire chief and placed in the hands of a third party at city hall or in some other level of government. But I suggest that the fire chief must be part of the labor-management relationship formula or the organization will eventually suffer for both labor and management. The fire chief does not need to become the chief negotiator; in fact, it is not desirable that the chief sit at the table and engage in confrontations with his own forces over individual items. Rather, the fire chief's job is to create an environment in the organization in which labor and management can be symbiotic. They may have two separate agendas, but they both recognize that they are part of the same organization and that the organism will not survive or grow unless they work together to achieve commonly held goals.

No one is promised, upon becoming fire chief, that dealing with labor-management relationships will be an easy task. Unfortunately, many of the skills, knowledge, and abilities that get an individual into the top job are not those that are required to establish good communications, articulate principles, develop open and honest communication, and establish organizational credibility. Enlightened fire chiefs are moving forward on this front, however, many fire departments now send their management and labor personnel to negotiation workshops as a team. When they return to sit across the table and debate the merits of respective issues, at least they know the ground rules and how to work together in a professional and productive manner.

If you take some time to examine the work of the various chief officers who have succeeded in maintaining positive labor-management relationships, you will be able to observe the dimensions of principles, honesty, and credibility at work in those circumstances. The degree to which these dimensions are adopted by you and your organization may have some bearing on the degree of satisfaction that you and the members of your organization obtain from your future labor-management relationships.

SELECTED READER ACTIVITIES

1. Prepare a list of individuals whom you already know you would like to have on a team. List their attributes and the reasons you would like to have them working for you.

2. Prepare a list of attributes that you would like to see in a hypothetical team to support you in the role of chief. Are there items on this list that are missing from the one developed in question 1?

3. Create your own definitions of the following terms: *confidentiality, trust, loyalty, honesty, credibility, accountability.* After you have prepared the definitions, apply these terms to individuals to whom you have been exposed both inside and outside your department.

4. Prepare a list of management practices that result in conflict with a labor organization. Discuss alternative strategies to avoid or overcome such conflicts.

5. Prepare a list of indicators that bad relationships exist between management and labor. Discuss strategies for avoiding these conditions.

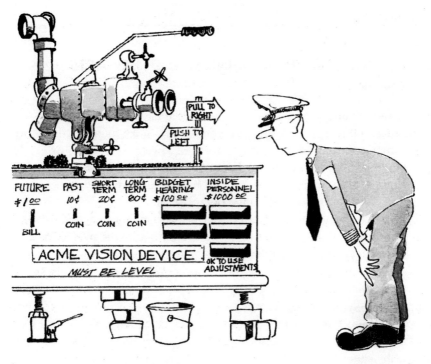

WHAT DOES THE FUTURE HOLD?

DOUG HUBERT RP

CHAPTER 13

Visioning:

Deciding What the Future Will Hold

The last two chapters have discussed getting control of yourself, and then of the team. The objective of this chapter is to describe the processes used in providing long-term, overall direction to the department. The emphasis in this chapter is on two levels of visioning: long-range goals and day-to-day accountability that achieves the vision.

> *Make visible what, without you, might perhaps never have been seen.*
>
> —Robert Bresson

PEARL DIVING: BRAINSTORMING AND OTHER ACTS OF CREATIVITY

What do oysters and ideas have in common? You have to shuck a lot of oysters in order to find a pearl. Worse yet, you have to find a lot of pearls before you can make an item of value in the form of jewelry. The same principle applies to ideas. You have to discard a lot of them as useless before you find a good one. Once you find a good one, it has to be linked to another good idea before it contributes much to the organization.

The fire service has often been criticized for lacking imagination. Fire chiefs have often been accused of lacking creativity. Our profession has often been attacked as "traditional." But before you react to these statements, I would like to make a few disclaimers. Yes, we are a highly structured occupation, and we tend to have a conservative bent. We have traditions and symbols that have not changed much over the last one hundred years. But is it true that the fire service lacks creativity? I think not. Our profession has made many more changes than many other professions have made in the same time frame.

Lacking imagination is not the most serious of our problems. Just look at the number of different ways we have figured out to load fire hose on our fire apparatus. In regard to the subject of fire itself, the amount of terminology and the extensive vocabulary we have developed from one end of the country to the other verges on the Tower of Babel. No, lacking imagination is not our problem.

The fire service, however, does have a problem. In this world of rapid change where new problems require creative solutions, we often waste a lot of resources on "reinventing the wheel." We do not lack creativity. What we lack is an adequate communications system to distribute creative solutions rapidly.

Our law enforcement peers have been dealing with this kind of problem for years. Criminals are mobile. Stolen property migrates. So, the law enforcement community has developed both databases and information exchange systems that are literally international. A police officer in Massachusetts can check out a vacationing traffic offender from California with the touch of a microphone key. And I am not talking about a Boston officer; I am talking about an officer from a town so small it has only one zip code. Law enforcement has a nationwide teletype system that allows for interchange of actual hard copy from a wide variety of databases. Criminal information and operational information can travel this electronic highway twenty-four hours a day.

The fire service's version is not quite that sophisticated. We have a great rumor network. Our conferences are chock full of opportunities to exchange business cards and promises of photocopies of programs, projects, and reports. But this technique has its limitations. You have to just happen to talk to someone who happens to raise an issue that fits with one of your problems. Coincidence and the "eureka syndrome" are far from being the kind of scientific communications system that we need.

The system used by the law enforcement community was paid for by the federal government. Our level of anticipation for a federal solution to this problem is best described in terms of "flights of fantasy" or wishful thinking. Nonetheless, you will need some kind of system to expose you to creative answers and solutions. If you have done any kind of assessment of the department as part of your transition, you probably already recognize that there are things you are going to have to change. There are a few possibilities that you should, as a fire professional, begin to explore and, if possible, exploit to help you find solutions. They are: development of a library of resources, the use of fax equipment, and the use of fire service Internet networks and web sites.

The most underutilized resources for information exchange in the United States fire service are the libraries located at the National Fire Academy, the International Association of Fire Chiefs (IAFC), and the National Institute of Standards and Technology (NIST). These libraries are the largest depositories of resource material and original research work available to fire administration personnel. Many people assume that they must have gone to the National Fire Academy in order to make use of its material. Many members of the IAFC have never accessed the Management Information Center (MIC). Many in the fire service have never even heard of NIST. These libraries are available to the entire fire service. One only has to make contact.

According to the library staff at the National Fire Academy, you may contact the library for location of a specific resource document that you know about but do not personally possess, for example, a federal study on smoke detector ordinances. The library can locate the document and arrange for you to acquire a copy. You also may contact the library for assistance with a study. The library will conduct a bibliographic search to identify materials that may relate to your topic. For example, if you were looking into physical fitness programs, the library could prepare a listing of various types of information—including books, studies, reports by National Fire Academy classes, and magazine articles—on that general topic. For further information on how to use this resource, contact:

Librarian
Federal Emergency Management Agency
National Fire Academy
16825 South Seton Avenue
Emmittsburg, MD 21727
(301) 447-6771

The International Association of Fire Chiefs also maintains a library. Although it is not as extensive as the Fire Academy's library, it does contain a great deal of information on projects that have been researched. For further information on the IAFC library contact:

International Association of Fire Chiefs
4025 Fair Ridge Drive
Fairfax, VA 22033-2868
(703) 273-0911

The National Institute of Standards and Technology maintains a library that focuses on the fire prevention engineering aspects of fire protection. For further information, contact:

National Institute of Standards and Technology
Building and Fire Research
Gaithersburg, MD 20899

The second thing that can help you facilitate the exchange of information in the fire service is the use of fax equipment, which was discussed in the last chapter. Many of the primary fire service organizations such as the IAFC, IAFF, ISFSI, and state fire chiefs' associations have fax machines in their main offices. If you have developed the capability to receive fax transmissions, you can literally exchange information on a real-time basis. For example, if the IAFC needs to send an emergency message to one of its committees, it can have hard copy into the hands of the committee members within minutes. Or a fire chief can call the library at the IAFC and ask for a copy of the recently adopted federal regulations on the hazardous materials "right to know" act or the new OSHA regulations regarding firefighter safety, and receive the actual legislation within minutes instead of waiting to read a synopsis six months later. But these information systems will not work unless you, the fire chief, have the equipment.

The initial cost of the fax machine is less than that for one pack-set radio, and that radio may get used only once every six months. The cost of transmission is the same as the cost of a long-distance telephone call, since the system uses the same telephone circuits. The advantage of fax equipment is that when you are researching a problem, the major organizations with information about solutions can communication with you. With this equipment, you make a quantum leap forward in information exchange.

The last idea is for a fire chief to develop his own external electronic mail service. Many, if not most, fire departments have computers in the office these days. Almost all of them are capable of being linked up with a modem to the Internet system. Instead of a rumor mill, you can have a network that is both responsive and easily accessed. The cost of operating e-mail is considerably less than that of dictating a letter, having it typed, and copied, and sending it through the mail where it is delayed for several days in transit. This chapter does not allow for an extensive overview of what is available on web sites and through e-mail—suffice it to say that it is extensive.

Imagine what you could do with the flow of information if you could send copies of letters and requests for political action to a thousand fire chiefs in less than a day's time. Imagine what you could accomplish in the way of influencing fire service response to legislative problems were you to have this capability. Are you presently able to respond to inquiries in a timely and cost-effective manner? It has always been interesting to me that a person can order flowers for his mother, wife, or a sick friend and have them delivered the same day, while we in the fire service often take weeks to transmit information on matters that are vital to fire chiefs. But like a trip of a thousand miles that does not get completed until you take the first step, having access to information requires that you act—that you take the first step. What do you think about it?

MISSION: PROBABLE, POSSIBLE, OR IMPOSSIBLE

Several years ago, there was a television show called *Mission Impossible*.[1] In the beginning of each show, the star, Mr. Phelps, received a tape recording that gave him an overview of a particular problem, generally one that bordered on being impossible to solve. The opening segment always ended with the tape self-destructing after stating something like, "And, Mr. Phelps, if you and your group undertake this mission and are caught and convicted, the government will disavow any claim of your existence." The team was supposed to take on a mission designed to save life, property, liberty, and equality, while the agency that empowered the team to take action stood back and disavowed the team's very existence. Sometimes the fire chief's situation feels exactly like this one.

---

[1] *Mission Impossible,* CBS, 1966–1973 and 1988–1989.

When a fire chief takes on a function, task, program, or activity that he believes is in the best interest of the fire protection agency and it somehow derails, it is not uncommon for local government to take a "hands off" position. More than one fire chief has gotten into in a great deal of difficulty by taking on something with an altruistic and honorable intent but running afoul of political realities in the community. The colloquialism of the day describes the situation as, "They got burned."

There are at least three steps that you can take to ensure that this does not happen to you. First and foremost is to make sure, during the self-assessment process, that the specific mission statement for your organization is appropriate. The second step consists of the development of a goals and objectives program in the organization to achieve the mission expressed in your mission statement. The third step is the development of an ongoing feedback mechanism that keeps the political and administrative staff to whom you report well informed.

First, let us look at the mission statement. You should have a mission statement already. In most cases, fire departments have written and published their mission statements, but some departments have accepted them by inference, and therein lies the problem. The mission statement is a declaration of intent. In other words, it clearly articulates what the organization intends to do in the scope of its activities. One of the first places you can go to find out what should be included in the mission statement is the enabling legislation that created the agency.

In my experience in working with fire departments, I have discovered that there are numerous fire departments that are involved in tasks and activities and even comprehensive programs that are clearly in violation of their basic enabling legislation. This may sound somewhat contradictory. Fire departments are created to handle emergencies, right? That is not necessarily so. In some cases, fire departments were founded to deal with a specific set of circumstances, for example, the formation of a volunteer fire department whose activities are restricted to fighting fires and explosions.

In reviewing the enabling legislation of several fire departments, I have discovered that, in some cases, the enabling legislation actually prohibited the department from engaging in certain kinds of programs. Over time, however, the limitation was forgotten by the administrative staff, and programs were incorporated. One might argue that if a program or activity has been funded through the budget process, it has been sanctioned. Not being an attorney, I cannot clearly defend or attack that position. In the review of some lawsuits, however, it has become clear to me that fire departments are vulnerable if they are involved in an activity with an exposure of liability. If liability exists because you are doing something outside of your original organizational intent, problems can occur. During transition, you should determine your status in regard to the enabling legislation.

Many fire departments have their mission statement nicely typeset and posted where everyone can see it. That demonstrates another important aspect of a mission statement. Everyone needs to know what it is—including the city council, city manager, board of supervisors, and any agency that is involved in supervising a fire protection program. With respect to the mission of a fire agency, there should be no surprises.

In discussing this issue with a variety of fire officers, I have been told that the statement of the department's mission has already been handled by public policy statement that the fire department is designed to save lives and property. But that same mission could be stated for law enforcement, public works, health departments, and so forth. Mission statements need to be specifically oriented around the agency and its resources. For example, the mission statement for one of my departments was as follows:

> The mission of the Fullerton Fire Department is to provide a range of programs and activities designed to protect the citizens of Fullerton from fires, medical emergencies and other sudden emergencies created by either man or nature.

The mission statement provides focus. It provides direction. It does provide some limitation, but on the other hand, it empowers the organization to accomplish certain things. In order to be more specific and to prevent involvement in activities that are contradictory to the mission statement, however, you have to take the next logical step: the formation of goals and objectives. This is one of the key responsibilities of a fire administrator. The topic of preparation of goals and objectives, and even the definitions of those two terms, is fodder for an entire semester's course. It is not the intent of this chapter to deal specifically with all the ramifications of the concept. Instead, I would like to focus certain distinctions regarding goals and objectives.

A goal is a strategic direction. A goal is something that you would like to see achieved at some point in the future. Essentially it is change-oriented. Using the sports analogy, in a game where the score is zero to zero, nobody has achieved a goal; and nobody wins until somebody scores. An organization's goals should state the conditions that should be achieved in the future if the organization is to win in terms of carrying out its mission statement. For example, in terms of a fire department's mission to protect people against fire, a major goal might be the development of a comprehensive fire prevention program to reduce the frequency of fire. Another major goal might be the establishment of a public education program to change people's attitudes toward fire. Or the goal might be simply to achieve an operations capability that would provide fire suppression operations within a certain time frame so that fire losses would be reduced.

Goals are somewhat open-ended in that if they could be easily achieved, they would already have been accomplished. A goal that is oriented around the status quo accomplishes little. Many fire departments, have however, taken the goals and objectives concept to the point where the goals and objectives have become more important than their completion. That is not what is being suggested in this chapter. As a matter of fact, goal programs that become little more than a paperwork mechanism rather than an exercise in reality are counterproductive. The goals are tied to the mission statement in the relationship between the direction of the organization and the levels of achievement in the organization.

Once again, we can relate to the sports analogy. A football team may have as its goal to win the Super Bowl. One of its goals might be to win more games than anyone else in the league. Unless the team keeps score, it does not know how well it is doing. In the fire service, if our job is to protect lives and property from fires and medical emergencies, then we should be setting up goals that let us know how well we are doing on those particular issues and in those particular activities.

Obviously goals have a long-term tenure. If the mission statement is in existence for an organization over its life span, then goals must have some sort of incremental evaluation process. Commonly, goals are divided into three categories: ten-year, five-year, and one-year goals. Ten-year goal statements, in most cases, deal with long-term strategic accomplishments, such as the building of facilities and the development of operational capability. Five-year goals tend to focus on the success or failure of specific programs. One-year goals tend to be focused more on activities of specific divisions and elements of the fire protection agency. During transition, one-year goals become very important.

Objectives are much more specific than goals. Objectives are those things we have to do in order to achieve the goals. They are more detailed with respect to who has to accomplish them, when they should be accomplished, how we go about measuring them, and who is responsible for their ultimate accomplishment. Objectives are of a much shorter time frame than goals. In some cases, an objective may cover the activities of only a given day, for example, fire service day, or a given week, for example, fire prevention week. On the other hand, some objectives might take an entire year, for example, the establishment of an officers' training program.

Both goals and objectives have to be articulated or published so that everyone is aware of their existence and has the ability to interact with them.

If goals and objectives are the best-kept secret in the organization, the possibility for derailment occurs once again. If an organization is attempting to accomplish something that is not in concert with the intent of the political entity, and especially if resources are being expended on a daily basis to achieve something that is ultimately discredited, the fire chief can be put into an uncomfortable position.

Goals and objectives should be reviewed, documented, and distributed in the organization in two fashions. First, the goals and objectives should be approved by the administrative agency that supervises the fire protection agency, and a consensus should be gathered with regard to the achievement of them. Second, the goals and objectives should be distributed downward in the organization so that people know the "game plan."

This leads us to our third and final element of preventing disavowment: the utilization of an ongoing feedback mechanism. Once goals and objectives have been established in the organization, it is the fire chief's responsibility to set up a device that will ensure that someone is working on the goals and objectives and that milestones are being achieved on an incremental basis. It is a truism, however, that a constant focus on the accomplishment of goals and objectives may actually be counterproductive. Probably the most frequent assessment required of such a program is a quarterly one. A certain amount of evaluation can be done merely by observation. If the document has been presented to all members of the organization, as various things get accomplished, it will not be necessary to ask for that information; it will be clearly available. When activities take an extended period of time, however, such as program development or facilities construction, it is appropriate for the chief officer to meet with his staff officers to assess the status of those activities.

In many cases, fire departments have linked goals and objectives programs with performance evaluations. This situation has its pros and cons. There is a tendency at times, when goals and objectives are linked to specific performance evaluations, to focus more on why things have not gotten accomplished instead of how they can get accomplished. There is no doubt that performance of an individual staff officer may be linked to the accomplishment of goals and objectives. However, one has to be careful not to mix tasks with specific job behaviors.

In any case, at the minimum, a semi-annual evaluation of the goals and objectives program is needed for adequate feedback. The feedback system must work two ways. As a fire chief, you need to keep your superior informed of how well the organization is doing, and the staff officers need to keep you informed about how well their divisions or elements are performing. If, for some reason, the material is not reviewed on anything less than an annual basis, it becomes increasingly important that it be linked with some other form of administrative policy and practice such as the preparation of the budget. Usually the accomplishment of goals and objectives costs something. Resources are discharged in the way of either purchases, personnel expenditures, or the allocation of funds.

The review of a goals and objectives program on only an annual basis has potential problems. Very simply stated, if we keep score only once a year, we may find that we are losing in the final hour. Almost all games in sports are broken into quarters, innings, sets, or some other division so that one has perspective on the score as it relates to the overall period of time set aside for the game. The same concept can be applied to a fire department. Dividing our yearly activities into months, quarters, or other incremental divisions as part of our feedback system is nothing more than another way of keeping score.

In the *Mission Impossible* television show, invariably Mr. Phelps and his crew succeeded. They were never disavowed, but they were never given credit for their successes either. In the final analysis, one of the reasons that a mission statement, goals and objectives, and a feedback system are essential to fire departments is credibility. To quote a statement made by IAFC past president Bob Ely during a board of directors' meeting, "We are what we are perceived to be." An organization that lacks focus lacks a game plan and lacks a means of achieving credit for its various accomplishments, ultimately losing credibility in the battle for public resources. An organization that has these mechanisms in place has a "can do" attitude that translates into "Mission: Possible and Highly Probable." How is your organization doing in this area?

IT'S NOT WHAT YOU DON'T KNOW THAT WILL KILL YOU

The sign read, "It's not what you don't know that hurts ya! It's what you know for damned sure that ain't so that will get you killed."

Overconfidence is a killer. It has humbled and destroyed individuals, organizations, armies, and nations. It is a characteristic of people who become so good at something that they think they have nothing new to learn. Overconfidence works for a while until you run into an adversary who does not care what you think you know, but who concentrates on what he does not know and beats you because he acquires a new skill that you lack.

The focus in this section is on preventing overconfidence in both individuals and organizations. Overconfidence is not a problem for people who are struggling; it is a trap for people who are doing well. It is quicksand for those who begin to believe that their past achievements will sustain them forever. If you are among the former, you still need to be aware of the trap, since there is a tendency to believe that once you have overcome adversity, you will always be able to use the same tools and techniques to retain your position. This belief is wrong. What you have done in the past has contributed only to your present achievement.

There are several things that you may already have in your own bag of tricks that give you the ability to win. They are: information and intelligence gathering, development of latent talent, focus on players in specific positions, and maintenance of a level playing field. Let us look at each of these elements more closely.

Information and intelligence consist of what you know, regardless of its source. It could be from experience, reading, education, training sessions, or whatever. It is your source of fact. Granted, facts sometimes coalesce into opinions, but initially most of us learn our facts raw. The more we understand the orderliness or relationships of facts, the more quickly information becomes an asset and contributes to our edge in competition.

Latent talent is the skill and ability that a person brings to the job that they have achieved. Latent talent can be natural, untutored, or even intuitive. Skills and abilities are basic strengths that we do not have to consciously call into place; they are just there. The development of your latent talent means that you focus upon those strengths.

Focusing on your playing position is evaluating your experience from where you are in the system. It is looking closely at the job, task, or responsibility that you and your staff have right now. A player's position places specific demands upon the mental and physical capabilities of that person. If a person's latent talent supports that, the person is in good stead. Different playing positions, therefore, draw upon a person's range of physical and mental competencies to widely divergent degrees.

The playing field has to do with the competition you are facing. You may be competing with people who have about the same amount of potential that you possess. They have the same kind of information and latent talents, and they are playing the same position. The field is level when this is the case—when, for example, amateurs are playing other amateurs, and professionals are playing professionals. All groups feel more comfortable in competing when the competition consists of peer groups.

All four of these elements come together when you pursue any form of achievement. Dropping the sports analogy for a moment, take stock of how these elements contribute to your confidence in what you are doing today in the fire service. If you know the facts, the rules, and the processes, and you have developed skills and abilities to perform well in the job that you are currently occupying, and you are doing it alongside a group of peers with similar expectations, you are probably fairly confident. It is a "feel good" situation. If this is not your situation, you may be feeling uncomfortable or even stressed.

Now, for the sake of discussion, let us elevate the first two elements— information and skill—but leave the last two, the playing field and the peer relationships, alone. It is easy to obtain more schooling and more practice and to get better in your current position than your peers. It is easy to get so good that you can begin to look down upon your peers. You may feel that you are better than them. Sometimes they will even admit it. More often, they will resent it. In either case, you can begin to see yourself as so good at something that you think you can do anything. Confidence swells based on that scenario.

What would happen if we changed the last two elements but kept the first two intact? In this scenario, you are asked to play a new position in a higher league but with only the information and talent that you have obtained in the minors. There are those who can make this transition, but they are rare birds. One of the best I can think of from the sports field was Babe Ruth. He had a talent for blasting home runs, but he was a pitcher who only got to play every few games. So the coaches made him an outfielder and a sports legend. Contrast his experience with that of the basketball star Michael Jordan who gave up an NBA career because he wanted to become a baseball player. He was a media anomaly for a few months and then faded in the sports world as a baseball player. He later returned to play again in the NBA as a superstar.

In the context of our profession, the four components are in play as we move from the rank of firefighter to the rank of fire chief. As we begin to function with more and more responsibility, we do not often talk about the fact that we can be programmed for potential problems. But people who have moved up in rank or changed from one size or type of department to another have failed. The reason for their failure was that they did not pay attention to the development of their skills and abilities for the entire range of tasks they had to face. I have witnessed this phenomenon hundreds of times, and in interviews with these individuals, I have heard them declare, "I just don't know what went wrong. I was really confident I could do the job."

And there is that word *confidence* again. An individual who fails may have been confident in his last job and, therefore, assured that he would be successful in the next one, but he did not assess why he was successful in his former job and what obstacles to success existed in the new one. What has worked in the past is not very important as it relates to the future. It is what must be used in the future that is essential for survival.

It is very difficult to convince people who are overconfident that they are too cocky. Pride, and even ego, makes identification of overconfidence in oneself difficult. We do not often recognize it when we look into the mirror.

I'LL SEE IT WHEN I BELIEVE IT

You may think that the title of this section is misquoted, after all, should not seeing precede believing? The answer is no. If you do not believe that something is there, then you cannot see it. Once you do believe that it is there, no matter how many different ways it is altered, you will still fill in the blanks to make sure that you see what you believe. This particular phenomena has been borne out in many psychological tests. An example is the test where you are shown various geometric objects that you are supposed to analyze. As soon as you are told what shape is within the object and believe that it is there, you can see it, but until that point, you cannot see it.

A similar phenomenon exists in the fire service, it is called level of service. Level of service means different things to different people, and those who believe it to be one thing cannot see it in any other form. This contradiction lies at the root of one of the most serious challenges facing the American fire service, that is, the creation of acceptable levels of service by the community you serve. It is also one of the most fundamental challenges that the fire chief of the future will face.

The creation of acceptable community-based levels of service is a complex topic with a simple title. There is a tendency to attempt to resolve the whole issue by trying to define a level of service that is the same for every community. That approach is doomed to failure for a very simple reason. Communities have many things in common, but they also have many dissimilarities. Interestingly enough, a lot of the commonalities have to do with the root of the American fire problem—people, places, and processes—while a lot of the wild cards in this system have to do with politics—performance and practicality.

First, let us define level of service. A level of service is nothing more than the enumeration of the amount of resources that has been set aside and devoted to a specific function. You cannot define a level of service in terms of outcome. All you can do is predict what you are going to have available when the service is required. For example, a level of service in the community can be defined as the number of fire stations, the number of fire apparatus, the staffing on that apparatus, and the ability to either redistribute those resources or concentrate those resources under a given set of circumstances. Level of service is one of the questions that should be addressed by a newly appointed fire chief during transition.

By providing a specific level of service, there is no guarantee that you will not burn buildings to the ground, lose lives of both civilians and fire-fighters in fires, or suffer financial setbacks to the community because of the loss of specific occupancy. You might call these the horns of a dilemma. You can describe a level of service by enumerating the availability of resources, but in most communities, three other factors come into play. The community has a political structure that is either for or against the deployment of those resources. The community has a level of expectation about what it wants those resources to do when you are required to use them. And, in the final analysis, those resources must be made available through either community commitment or a budgetary resource allocation process.

Some of you may be aware of the recent attempt to quantify this process through the development of a "fire protection master plan." Although this work is still very meaningful to the fire service, it has been, for the most part, ignored by communities in their development of acceptable levels of service. One of the stated reasons for this is that the creation of a master plan is a laborious process. That does not have to be the case, however. The concept is very basic: Most communities consist of people, places, and processes. People cause fires, the places where they live and work are threatened by risk, and what they do within those buildings results in a frequency and severity of fire. The more that top fire officials know about these three factors in their own communities, the more likely they are to be able to articulate the issues to the uninformed.

Although these three factors have been defined in the master planning process, they are often overlooked in the development of commonly accepted service levels. If we do not know what our community fire problem is, then who does? The bridge between the problem and the provision of service is probably one of the most complicated bridges to cross. In order to cross it, the top fire official needs to have a firm grasp on the capabilities of his resources to deal with the problem. This requires an in-depth analysis of such things as critical tasks, a measurement of efficiency and effectiveness of various programs, and a set of operational parameters that are acceptable to the community.

An example of an operational parameter is response time. Many fire departments have used response time as part of their explanation of service level, but they have failed to realize its weakness when used inappropriately. For example, many departments state that they have an "average response time of five minutes or less," which sounds good until you realize that an average five-minute response time means that fifty percent of the time it is far greater than that. Such statements raise the level of expectation in the community by making it sound as if five minutes is the normal response rate, and then a person who receives a response in seven or eight minutes feels cheated. A better approach would be to state that response time is "five minutes or less eighty percent of the time." The statistical difference between these two definitions is very significant.

So, how do you deal with the level of service issue? Do your homework! Use the self-assessment technique discussed in Chapter 10. A jurisdiction is a fixed geographical boundary with a road network and an occupancy of people. Measure the community in every way possible. Then take those measurements to the drawing board and determine your community's level of service right now. You must start with the current level of service, but you cannot improve it unless you can describe it.

By using this approach, you might find, for example, that your response time is six or seven minutes, not five minutes or less, eighty percent of the time. You might find community support for reducing that response time, especially when it is placed in the context of the need for life support. The message needs to be taken to civic groups, the media, city council, fire board administrators, and anyone who will sit and listen to your information about the level of service that currently exists. It is an educational process. It is also a community definition process. One of the problems the fire service has from time to time is believing that we know what is best for the community. However, as you may remember from our earlier discussion about politics, performance, and practicality, if we cannot convince others that this is what we need to accomplish, we will never get the resources to accomplish it.

Granted, it would be a lot easier on all of us if there were one simple formula that could be applied everywhere. The demographics of the American fire problem, however, make this extremely difficult. We are not in this dilemma alone; many business entities face similar problems. For example, have you ever considered the kinds of management decisions that must be made in the placement of franchise restaurants? These restaurants must be placed where they will make a profit. Therefore, long-range planners frequently look at such things as use rate, availability of traffic circulation network, and the class structure of the community. These managers face level of service questions that are similar to the ones we face, and they solve them through analysis.

The final step in dealing with acceptable community levels of service is to get them institutionalized. There are many places where this can occur. The first of these is in a budget document. Once you have created an acceptable level of service and placed it into a budget document, you must then follow up to make sure you measure it, year in and year out. Failure to monitor a level of service makes it remarkably easy for subsequent administrations to question the validity of it.

Merely placing a fire station in the middle of a neighborhood does not ensure that fire station's survival if the neighborhood changes. Placing a fire company in service and then acting as though anything goes with its performance ignores the fact that, in most communities, all other service levels are in a constant state of flux. It is incumbent upon the fire chief to manage the level of service and the fire problem concurrently. Allowing one to be out of balance with the other usually means that there will be either a series of catastrophic losses in the community or a complete and total erosion of the resources deployed to protect the community.

Creating an acceptable level of service for our community is based on a comprehensive understanding of what level of service the community wants. That level of service is then reinforced, as often as possible, as being a community level of service rather than the fire department's level of service. During transition, you need to develop a good perspective on what you see and what you believe about the organization in regard to its current and potential level of service.

SELECTED READER ACTIVITIES

1. After reviewing the self-assessment document from the department, what changes would you make to the mission statement? What goals would you establish, and what objectives would need to be accomplished?

2. Develop a budget. Join the IAFC, your state fire chiefs' association, and your county fire chiefs' association.

3. Prepare a list of three topics about which you know very little but which are pertinent to your department's growth and development. Contact the IAFC library and determine whether packets of information are available on those topics.

4. Prepare a presentation on the topic of level of service and response time to deliver to the authority having jurisdiction.

5. Review the community's strategic or master plan to determine where the fire agency fits into the general framework of the community's future.

CHAPTER 14

The Assistant Chiefs:

Your Spouse and Your Secretary

The objective in this chapter is to explore the need for developing and sustaining understanding with two key people in the life of the fire chief: the chief's spouse or other significant person and the chief's executive secretary.

> *The best index to a man's character is (a) how he treats people who can't do him any good, and (b) how he treats people who can't fight back.*
>
> —*Abigail Van Buren*

TILL DEATH DO US PART

This is one of the few times that I have written something that I think needs a warning label. Caution: Reading this chapter may be painful. It may bring back some unpleasant memories. It may make us think about something that to date we in the fire service have not faced, and probably do not want to face. That something is called divorce. It happens to people all the time. In the fire service, it occurs far too frequently as both a personal and an organizational problem. For fire chiefs, it is a serious consideration.

Divorce seems to happen often to individuals as they move higher and higher in rank or responsibility in the fire service. Although I do not have a scientific database to support my hypothesis on this issue, I have a certain amount of direct and indirect experience. Early in my career, I went through a divorce, and many of my cohorts have done likewise. I have seen hundreds of employees, friends, and professional acquaintances experience the pain and anxiety of this process. Once I was involved in an after-hours bull session where I revealed some of my personal experiences, and a dozen or so fire chiefs who were present all indicated that they had been divorced, some more than one time.

Has it happened to you? If so, I am sorry. Could it happen to you? If you say no, I am hopeful for you but concerned that you may be more vulnerable than you think. If you admit that it could happen to you, perhaps you are the kind of person for whom this chapter has been written. Those of you who are married, who want to stay that way, and who are willing to pay attention to some new possibilities might acquire a few pointers here.

The reason for writing this chapter is twofold. The first is the empathy I have for those of you who have experienced the trial and the losses that almost always emerge from this situation. If you have already experienced divorce, there is nothing that can be done to avoid it. But the second reason for this chapter is to offer you a chance to avoid divorce after you have made the leap to fire chief. Divorce happens at many different times in career development, but the potential increases with the increased pressures and stress of the chief's job.

First, let us deal with the empathy part. Divorce is seldom a neat, clean, tidy bifurcation of two people's property. It commonly affects children, friends, family, and even your professional capacity. Many individuals who experience divorce lose all of the things that they have spent years accumulating, including the quality of their own life. They frequently lose affection of both family and friends, including children, and sometimes they even lose their retirement benefits.

When I was going through my divorce, I thought that I was handling it fairly well. Years later, my superior told me that he knew when things were deteriorating at home because I had changed direction at work. I might have done a good job of fooling myself, but I did not fool him.

First we need to dispose of the two most common reactions that I have heard when discussing the issue of divorce in the fire service: (1) I am glad that I got a divorce because (he) (she) was not the right person, and (2) My second time around I found the right (man) (woman) to live with, and I am happy again. With regard to incompatibility, I will not dispute the fact that some people never should have been married to one another in the first place. There are amicable separations based on mutually exclusive reasons. Some divorces, however, are so messy because the people are still in love with each other but can no longer stand one another; they do not know how to deal with each other, so they attempt to accomplish that task by ripping themselves apart. The messier the divorce, the more likely it is that passion and affections are being misrepresented as jealousy and conflict.

Sometimes we divorce the right person for all the wrong reasons. Second marriages are often better marriages because people have learned to be better spouses. It is not that the second marriage is better; it is just that the second marriage works better. Many of us who are divorced probably have ex-spouses who are now remarried, and their second marriages are probably working better as well. People mature and grow from adversity. Often we help overcome our losses by learning to prevent them in the future.

But there is a point of diminishing return. My own father was married so many times that we stopped keeping track. I often joke that every time I wrote home, I started the letter with "Dear Dad and For Whom It May Concern." The point here is that marriages and divorces are conscious acts that materially affect our own self-esteem and stability. I would not criticize anyone with respect to their past or future considerations on matrimony because that is a personal decision, but I would like to offer some insight and a possible resource that may help you to make good faith efforts in maintaining the relationships that you have right now.

The resource is a book entitled *Men Are from Mars, Women Are from Venus.*[1] The author of the book is John Gray. I had a chance to read this book while I was on vacation. Originally, I purchased the book as a sort of joke. As I read through the book, however, I was struck by the number of times I said to myself, I have been there and I have done that. First I just scanned the book quickly to see if I could find things to justify my own behavior, but I felt compelled to go back and read the book more closely.

I am not going to tell you about the book in detail; this is not a book review. My focus is on the book's potential for people who have a lot of pressure upon them to maintain matrimonial relationships while they are simultaneously dealing with stress and strain on the job. In other words, this book is for you, the fire chief. Gray is onto something that can help all of us who operate in highly stressful jobs. His approach is similar to that of other self-improvement books. He has placed the wisdom behind a catchy slogan. I would almost risk a paycheck that most firefighters would never pick up this book merely because of its title, which is catchy but has that touchy-feely sound that we tend to avoid. If you can get over this reaction, however, you should buy the book and read it. It will do you and your spouse a world of good in looking toward things in the future.

Gray's concepts regarding relationships move beyond husbands and wives to that complex matrix of male and female relationships we all have that involve affection—father-daughter, mother-son, parents-in-law, and so on. These relationships sometimes go sour. Some of us have lost touch with close friends after some spat over a trivial matter. Gray's theory is very basic, and yet his anecdotes are almost all universal. Once or twice I had the feeling of knowing the outcome of a paragraph after reading the first two or three sentences because I had experienced the situation he was describing myself. This book is based on real world versions; his examples are played out daily in your lives and in those of your family and friends.

I would advise you to look up at least two of the techniques that Gray exposes: the first is the letter writing exercise, and the second is called the coulda, woulda, shoulda and concerns the ways in which we ask for help. Both of these are classic techniques. The former approach I had actually practiced many times without knowing it was a technique; I have many letters that I have written to people but never mailed. The latter approach has to do with asking for support in ways in which you can accomplish your goals by reducing stress in the asking phase. This section relates to

---

[1] John Gray, *Men Are from Mars, Women Are from Venus* (New York: HarperCollins, 1992).

communications with our spouses but it also applies to communications with our staff and others who support us.

Would you like to see things get better in your relationships? Buy this book and read it. Give it to someone you care for and let that person read it; then, sit down with that person and talk about it. I am not saying that this one book will revolutionize your life, but it will shed light on things that have troubled you in the past. This book is not a complete solution to complex problems, but it is one of the best explanations I have ever read on how we interact closely with those we care for the most.

I am reminded of a short story by O. Henry that had a profound effect upon me as a teenager. It is called "Gift of the Magi." In this story, two people give away their most treasured possessions in order to give something to each other. The fatal flaw is that what they give up to purchase the gifts is linked to the gifts that they each receive, so the gifts are useless. What they gave away was priceless. One of the greatest gifts you can give to someone that you care about is to understand them. Gray offers you an opportunity to become much more blessed by giving than by receiving. He has also authored another book with a discussion of second marriages, and it may be an excellent companion piece for your consideration.

THE ASSISTANT CHIEF

Suppose I were to ask you, "Who are the key members of your staff?" You would most likely respond with a list of names prefaced by rank: Captain So-and-So, Battalion Chief So-and-So, and Assistant Chief So-and-So. Or you might respond with functional titles: Training Officer, Fire Marshal, and Operations Chief. Although these *are* key members of your staff, there is another member who is almost equally important: the chief's secretary.

The position of chief's secretary is sometimes so important in an organization that it is given the informal label of "Assistant Chief." There are legitimate reasons why this might occur. In the first place, the secretary to the fire chief is right next to the information source. She is the person who drafts the policies, practices, procedures, communications, and letters of discipline. The person in this position can be either a very powerful person or a disruptive influence in the organization. In our memories, we can probably find examples of both.

When the chief utilizes the secretary's position in a proper fashion, not only will the secretary make the chief look good, but the secretary will improve in her position as well. Even though our secretaries will probably never wear gold trumpets and command forces at the scene of an emergency, we need to give due consideration to their influence in our organizations.

Failure to do so has resulted in many a chief officer stubbing his toe on something that should not have been a serious problem.

The secretary whose first loyalty is to the boss does not leak information that should not be distributed in an organization until the appropriate time. A secretary who is firmly committed to the same goals and objectives as the boss does not allow the boss to forget deadlines, does not fail to make appointments on time, and does not make serious errors when turning in paperwork to the boss's superiors. A fire chief has a right to expect certain things from the secretary, but it is a two-way street. The secretary also has a right to expect certain things from the fire chief. For purposes of clarification, let us divide these rights of the fire chief and the secretary into three categories: expectations, competencies, and courtesies.

Fire chiefs tend to take their secretaries for granted, and secretaries tend to take their bosses a bit too seriously. One of the most important things for chief officers and their personal secretaries to achieve is comprehension of what each expects of the other. A chief and his secretary need to be in agreement about the manner in which the work is accomplished in the organization and the manner in which they will communicate with one another. For example, some chiefs go through time management courses and establish an A, B, C priority system for how to handle their work, but then neglect to tell their secretaries which items are A, B, or C. The chiefs process material, place it into the out basket, and expect the secretaries to use a crystal ball to figure out which items are most important. The most critical understanding to be achieved between the chief officer and the secretary is what the priorities are and how objectives are going to be met by the two of them.

Other expectations that might need to be defined by the chief include such things as an agreement about what the secretary can do without direction and what requires a specific decision by the chief. Chief officers multiply their daily work hours by delegating many paperwork tasks to their secretaries. In some cases, secretaries are authorized to prepare rough drafts of standard responses instead of waiting for the chief to dictate those communications.

The matter of telephone protocols and courtesies are also part of the expectations field. For example, if a chief officer has a predisposition toward telephone numbers and names being handled in a certain fashion, the secretary should be told.

Conversely, the secretary should be given the opportunity to state her expectations of the chief. In a conversation with a highly regarded secretary, I asked her to state the one attribute that she would most desire in her chief. She stated, "I want my boss to support me in front of other members of the staff—to not make me the butt of any mistakes or problems, and to let the staff know that I am a part of the team and am regarded as such."

It is difficult for some people to understand that their subordinates have expectations of them. If you expect loyalty from your subordinates, you have to be willing to give them loyalty. If you expect honesty, you have to give honesty. If you expect your secretary to set priorities, you must provide your own priority system. The conversation it takes to develop these expectations frequently cannot be accomplished in a single session; rather, the expectations build and are clarified over a period of time with give and take between both people.

Competencies are another matter. A fire chief has the right to expect certain minimum capabilities from a person serving as a key staff member. While many of these competencies are clearly stated in the job description, they need to be monitored and assessed during a performance evaluation of the secretary. Among these competencies are such things as typing, transcription and editing skills, interpersonal behaviors, and the operation of office equipment. Most importantly, however, the chief needs to address with the secretary what level of competency is acceptable. Although this may appear to be part of the expectations discussion, there is a clear-cut distinction. Expectations can be stated at the outset of a working relationship between two people, while competencies may take a considerable period of time to evolve. In my personal experience, I worked with a secretary where we had very clear-cut expectations from the beginning, but she did not achieve a competency level that I found to be satisfactory for nearly two years. It is not beyond reason for a chief officer to expect the secretary to master certain skills to the point that they require minimum attention from the chief in order to achieve the organization's objectives.

But what about the competencies of the chief? Does the secretary have the right to expect the boss to know how to do the chief's job? This is a bit more tricky. For one thing, if you have been placed in charge of an organization, there is the assumption that you must know how to do something right or you would not have been placed there. Unfortunately, however, this is not always the case. Many individuals have migrated up through the rank structure in an organization but lack basic skills in dealing with clerical support staff. In many organizations, individuals at the lower levels send everything that goes to a secretary through some sort of screening process. When you are given the responsibility of supervising the secretary, your relationship changes, and the types of skills required are different from those required to merely utilize the secretarial staff.

Among the competencies that a secretary looks for in a chief are that person's ability to plan work, organize material, perform basic dictation and editing, and understand the problems the secretarial staff may have in implementing the wishes of the chief. The chief officer must master the skills that will improve the operating efficiency of the secretary. For example, I once talked to a secretary who told me that her boss insisted upon writing every single piece of paper in the office in longhand, despite the fact that the handwriting of the chief officer (not unlike my own) was almost illegible. The chief officer refused to learn to use dictation equipment on the assumption that it was dehumanizing and made it difficult "to think." The net result of that lack of distinctive competency was that the organization operated at a snail's pace.

Although a chief officer will probably not have a class on dictating skills included as a part of a chief officer certification program, it is one of the competencies that should be acquired in working out a professional relationship with a secretary. It is common for top-notch secretaries from other industries to come into employment opportunities in the fire service as a result of retirement or deaths in the family after having worked for attorneys, doctors, and so on. It is not humorous for the fire service to be considered "second-class citizens" as office managers.

Courtesy, while it is the last of the rights being discussed, is among the most important. The exchange of courtesies between the chief and the secretary should not be taken lightly. In many cases, what is classified as a "perk" of a job, can become a matter of conflict in the professional exchange between the chief officer and the secretary. For example, the act of sending the secretary for a cup of coffee for the fire chief may be a status symbol for the chief, but it is demeaning to the secretary.

One of the first and most common courtesies is mutual respect. This is essentially nothing more than demonstrating on a daily basis that you share mutual regard for each other's functions and position. One of the things most secretaries are extremely sensitive about is knowing the whereabouts of the individual to whom they are responsible. Fire chiefs should keep their secretaries apprised of how to reach them and when they expect to return to the office if they expect their secretaries to be able to respond to inquiries regarding the chief's activities. Sometimes such common courtesies as staying out of the filing system for which the secretary is responsible have rewards.

Whenever you are referring to your secretary, the tone and implication in your voice should always be in a positive vein. Even when your secretary makes an error or does not quite meet your expectations, she will certainly not get any better with public criticism. In the vein of the "one-minute manager," it is far better to find reasons to praise your secretary to reinforce good behavior than it is to criticize the mistakes that she makes.

Your secretary may not be the Assistant Chief, but she is an assistant to the chief. If you recognize your secretary's abilities and, at the same time, your own limitations, there is the possibility that the two of you can grow together. A good secretary makes even a mediocre chief look good. A bad secretary can make even the most outstanding chief look like a bozo. The rapport between you and your secretary is almost entirely up to you.

SELECTED READER ACTIVITIES

1. Develop a master calendar of personal events for which you should place your spouse as a number one priority, for example, anniversaries, birthdays, and other special events. Put them on your calendar at work.

2. Keep your spouse informed about what is going on by sharing both information and access to that information.

3. If you have not experienced domestic problems as a result of career development, celebrate that fact itself by telling your partner how much you appreciate it.

4. Conduct a personal interview with your secretary and discuss both of your points of view on what makes the partnership work well.

5 Never, never forget National Secretaries Week. Note your secretary's birthday, anniversary, and other personal events in your calendar as well.

EAGLE TURTLE OR TIGER?

DOUG HUBERT 9.0

CHAPTER 15

Are You an Eagle, a Tiger, or a Turtle?

How Are You Going to Lead the Organization?

This chapter focuses on how the chief, who is now responsible for everything that happens in the organization, provides direction. The chief can no longer blame the preceding administration for what happens. What does the chief do to keep the organization moving forward? The emphasis in this chapter is on appropriate ways of identifying, and some considerations in dealing with, the performance levels of department members who are under the chief's guidance.

> *Change does not mean progress, nor does it mean decline. It only means movement from what was old to what is new. Change is neither good nor bad, unless it is defined to be so by a person or*

institution that is affected by it. Change not only means different, but it also means direction.

—Ronny Coleman[1]

THE ODYSSEY OF IMPLEMENTATION

A large number of the early Greek myths have to do with the gods taking journeys. In order for these heroes to achieve monumental reputations, they had to overcome a tremendous number of obstacles. They had to deal with many adversaries in order to achieve their objectives. Having done so, they were recognized as having their place among the legends of their time. The premise was that once an individual began some kind of journey, an insurmountable challenge would test both his physical and mental capabilities. Once he had achieved the goal, of course, the world was a better place because of it.

If you take a close look at the stories of the journeys taken by mythological gods, you will notice something unique. The challenge is usually thrown down in the first few paragraphs of the story, and the actual conquest and achievement are almost always in the last few paragraphs of the story. The main emphasis throughout the epic is on the journey itself. An "odyssey" often involves taking right turns and left turns, stepping back, and, in some cases, making sweeping changes in the direction of the traveler's journey in order to achieve the overall goal. The odyssey itself is the story.

Taking that concept forward a couple of thousand years, the individual who advocates change in a fire agency faces a journey of a similar nature. Advocating change is often based upon meeting a new challenge. It involves overcoming conflicting points of view, obstacles, adversaries, and resistance to change. Once a chief has achieved his goal in the form of implementation, he may believe he can sit back and rest upon his laurels, with society thinking that he has achieved something of great significance. It seldom happens. Success is not always a satisfactory experience.

There is a lesson in the myths for fire service individuals who are trying to change the system. We should recognize that the challenge that is given to us has probably been given to many other people who came before us. In many cases, others chose not to look at the challenge—they turned a blind

---

[1] From a speech entitled "The Challenge of Change" given at a California Fire Chiefs Association conference, Newport Beach, 1997.

eye to it. Those individuals never got the chance to begin the odyssey. There were also individuals who began the journey but, for some reason, failed to meet the tests and were subsequently removed without recognition.

There are some things in this journey concept that must be part of your mind-set if you choose to take on major challenges in the fire service. One of the first observations is that whenever you strike out on a journey, you had better have a specific goal in mind or it will be easy to get lost in the wilderness. Too many people advocate change in the fire service without addressing a specific problem. Those people who choose to tackle projects and programs just to make change for the sake of change are often doomed to a high level of frustration, if not outright anger, over the fact that there are those who will prevent them from achieving change for change's sake.

The next thing you must remember is that whenever you begin a journey, there will be three groups of people about whom you need to be concerned. First, there are those who formed the support mechanism that helped us to become what we are. We have discussed them in previous chapters. Individuals who undertake an odyssey of change cannot get so removed from their basic value system that they forget where they came from. They must still be able to relate to family and friends. Second, in the Greek myths, the hereo was always accompanied by a group of companions. In a modern-day context, this group consists of our peers. It is made up of our professional friends who agree that our concepts are worth fighting for and who are willing to accompany us on the journey in spite of the challenges we are about to face. And then, of course, we must recognize that there are those who simply do not want us to achieve what it is we are attempting to achieve. They are the ogres and monsters of the unknown, but in many cases, they are equally as powerful in their attempts to achieve things.

Once again taking lessons from our mythological ancestors, you will note that in almost all of the tales alluding to heroes and legends, the heroes win because of wiliness rather than brute strength. The use of force in achieving objectives along a journey often results in more force being applied to resist those achievements. Whenever we begin to advocate change in the fire service, the thought process, the planning process, and the ability to conceptualize solutions are much more important than the mere ability to force a solution upon others. This is where leadership comes in. I have seen both winners and losers who have tried to create change through a process of direct confrontation. On balance, it appears that when people win through sheer use of power, it is usually a temporary set of circumstances, for when the power swings in another direction, many of the things that were accomplished are reversed.

Not unlike many of the mythological journeys undertaken by Greek heroes, almost all modern-day journeys have resting places and periods of respite along the way. The lesson to be learned by this is that no matter how severe the challenge or how long the journey, one of the things we must do is to pace ourselves and make sure that we are not mentally and physically exhausted when the final challenge arrives. An individual who is exhausted may find himself vulnerable at the point when action must be taken to bring about the ultimate solution to a particular problem. What you have to learn in the fire service is to make haste slowly and to enjoy the journey regardless of the number of conflicts you have faced in the past or that you see in the immediate future. The ability of a chief officer to sit back and relax once in a while even in the midst of a tumultuous process of change is an extremely valuable attribute.

In almost all of the Greek odyssey, the individuals who go on the journeys learn something about themselves. They grow in mental stature when they are facing stronger adversaries. The lesson to be learned here is that when we are undertaking change in the fire service and we find ourselves in conflict, sometimes with more enemies than allies, it is a time for us to look inside and determine what kind of inner strength we can bring to bear to the situation in order to suffer through the problem.

If we strip away the metaphor of the journey, the reality behind this discussion involves proposing changes to deal with the complex issues facing the fire service. Some of them are regulatory, such as sprinkler and fire alarm ordinances. Some of them are operational, such as the creation of or change in staffing patterns, fire station distribution, and training programs. Almost any form of change in the fire service is going to be difficult. We have a two hundred year old track record of both resisting and yet coping with change. If you visit a fire museum, you will see that all of the things that are worthwhile for the fire service have been adopted in time. All change has gone through the same process. In the beginning, the agent of that change (the fire chief) has undergone a difficult and tortuous journey. Nonetheless, someone has to make sure that changes will ultimately be implemented.

Do not begin the journey until you have contemplated the consequences of it. And do not try to introduce change in the fire service with the belief that the end will be immediately in sight. Take the perspective that once you have risen to the challenge, you are going to engage in a series of processes that will take you in many different directions. If you keep your eye on the compass and bear down on your goals, you will find the journey to be an enjoyable and rewarding experience. And you will gain even more from the adventure through your interpretations of it as the story is retold.

LEAD IS A VERB

The words in our language are either active or passive. Granted, English teachers will give you all sorts of definitions such as nouns, verbs, adjectives, adverbs, and dangling participles. But let us talk about basic language. Words are listened to or they are ignored. Words have impact or they miss their mark. In the final analysis, almost all of us choose to make words into whatever we want them to be. It follows that it is our own interpretations of words that give us the incentive to use them in communication. For example, an inflammatory word, even a profane one, when used in one context, can almost be a description of friendship.

Let us take a look at the simple word *lead*. You may recall a cartoon that was popular fifteen or twenty years ago that showed a little green man from Mars talking to some sort of mechanical device, and the caption read, "Take me to your leader!" Just what does that mean? Does it mean the person in charge? There are a lot of people in charge who do not provide leadership. Is it a position? There are many people in positions of power who do not provide leadership.

In the fire service, we often use the terms *command, control, administrative skill, management responsibility,* and even *rank* and *position* as synonyms for leadership. The title *leader* sometimes describes a person, but it does not describe what the person does.

Lead is an action word. One cannot provide leadership to an organization without taking action. It is mutually exclusive to maintain the status quo and then lay claim to being the leader. But one also has to take into consideration the direction and speed of the organization's movement: its mission, its goals and objectives, its action plans. We have already discussed the need to develop these in previous chapters. Now it is time to act upon them.

Action without purpose is simply frenetic. Purpose without action is stagnation. When I think of a person who has a leadership position but does not have a focus, I am reminded of a quote in I once saw in a fire chief's office: "Having lost sight of our initial goals and objectives, we have now doubled our efforts."

So, are you a leader who leads, or are you a leader in name only? Equally as important is the question: Are your actions reflected in the behavior of others in your organization? Is leadership being exhibited by your subordinate fire officers?

In order to lead, you must be committed to action. Furthermore, that action must result in some kind of change or you are not leading. The concept of leadership is valueless. There are bad leaders, and there are changes that occur in organizations that are counterproductive That is why there can be conflict even when leadership is exercised. A classic example of this is our political system. The individuals who rise to leadership roles in political parties are masters at challenging the status quo. Their primary role is aimed at throwing out the rascals from the other party.

We must also recognize that there are levels of leadership. Anytime there is a gathering of more than two people, it is conceivable that one of those persons will seize the day and take responsibility for the other. We tend to talk of leaders as if they exist only at the national or international level, but there is leadership exhibited across the spectrum. It can range from the leadership of the fire company on the opposite side of your town to the leadership of your regional, state, and national fire service organizations.

This chapter presents a litmus test of leadership. You will have to form your own conclusion about whether you are satisfied with your performance in a leadership capacity. The litmus test that I propose has only three criteria. You can apply them to yourself or you can use them to draw conclusions about others. Obviously, there may be other criteria that you can generate, but these three are easy to remember:

1. Are the goals to be achieved by the leader different from the status quo?
2. Is the leadership function embodied in only one person, or is there a followership?
3. If the leader were to be removed, would the cause go on?

Each of these criteria has implications. The first one, for example, demonstrates that leaders do not go in circles; they go from one place to another. If you are not challenging the status quo, then you are not leading. If you are not improving something, creating something, or advocating something, the first test of leadership has not been satisfied.

The second criterion is highly personalized. If others do not begin to rally around the change that is being advocated, personal power may sustain it for a short period of time, but as soon as the leader is gone and a new leader assumes power, the cause takes off in a different direction. Nowhere is this more evident in an organizational context than in the field of athletics. Dormant athletic teams have been known to become world champions with the change of the person at the head. The athletes have not changed— the game plan has changed. There are those who will not want to go along with change; they are the resistors. But leaders need followers. Do the members of your organization contribute to the actions of the leader?

Our third criterion sounds like the second one in that it talks about the absence of the leader. The significance here, though, is the emphasis on the cause becoming more important than the leader. Probably the best way of characterizing this phenomenon is to classify leaders as having created movements. Such things as democracy, religion, and perhaps even politics are examples of areas in which causes have gone off in a particular direction and have continued far beyond the lifetimes of their creators. Some of the contemporary changes in the fire service that fall into this category are the trends in training and education, the increased emphasis on the Incident Command System, and the increased use of built-in fire protection systems like sprinklers and fire alarms. New leaders can take temporary control of movements as long as they do not violate the perceptions of the followers.

If we place these three criteria in the context of an individual organization, we can see a microcosm that contains all three of them. First, if an organization is not going somewhere, it is not being led. Second, if the leader is changed and the organization takes off at a ninety-degree angle from its previous direction, *one* of the *two* individuals was not providing enough leadership. Lastly, if a fire service organization stops functioning when those at the helm leave, they were not providing leadership.

There are different degrees of leadership. Some people provide just enough leadership to get an organization moving, and then they rest for a while. Others function in an extremely high-energy state. These individuals consume a great deal of energy and are candidates for burnout.

When an organization is being led well, the members of the organization can literally feel it. Lack of leadership is felt, too, but in a different way. The sense of self-confidence and self-worth in an organization and the momentum of the organization are functions of the amount of energy devoted to moving the organization, whether that energy comes from the top level, middle management, or even the company level. Some people call it morale. Another characteristic of most effective leaders, which contributes to that morale, is that they tend to be *for* something instead of *against* something. The application of all this energy is for a positive outcome as opposed to a negative one.

What we are talking about here is not a science, but an art. The cliché about art mentioned earlier applies here: not everybody can define good art, but everyone seems to recognize it when they see it. If you begin to ask yourself the questions presented here and become a little more critical of your own performance, perhaps these questions will get you some new answers.

SURVIVAL OF THE FITTEST: EVOLUTION OF STYLE

Darwin was not the first person to think of the concept of evolution, but he was the first to write it down and publish it. The firestorm of controversy that arose from that publication gave him worldwide recognition. The fact is that several other people were talking about the same concept, but Darwin managed to put it into writing first. Although this is an oversimplification, Darwin's basic theory was: Organisms that adapt survive. Organisms that do not adapt become extinct.

When we look at fossil records, we have clear-cut evidence that this lesson about evolution has been taught over and over again to organisms ranging from microscopic to humongous. Power and energy, even ferocity and aggressiveness, have not prevented certain organisms from disappearing because of their inability to adapt to a changing environment. What does that mean to you? In the simplest of terms, the fire service is being forced to adapt, and there are some dangerous consequences of failure to do so. One consequence of that failure is to become extinct. Mother nature, in spite of her cynical survival-of-the-fittest approach, often allows remnants of a genus to survive to live among a new array of organisms. In terms of the future of the fire service, if we do not adapt but still survive, where will we exist in the overall scheme of things?

This topic is not new. There have been references to dinosaurs and such in the fire service for some time. This discussion, however, is becoming more relevant each day because the fire service is currently facing one of its greatest challenges in perhaps fifty years. The issue is, how can we maintain our equilibrium in a time of economic instability? This is not a rhetorical question. One only has to look closely at the budgetary statistics from many cities to realize that erosion is occurring. Departments are, in many cases, being asked to make cutbacks. Others are being forced into restructuring. There are cases where departments arc growing, but one has to be careful to distinguish true growth, that is, improvement in the level of service, from mere expansion to maintain a level of service as a community grows upward or outward.

There are too many variables for me to attempt to discuss the kinds of adaptations that must be made by individual fire agencies to survive these threats, but there is one observation I can make about our overall actions: We must learn to adapt. Failure to do so may be fatal. When conditions change fast, we must adapt faster. If the fire service refuses to adapt to a changing environment, the price will not be paid by this generation; it will be paid over the next twenty-five to fifty years when public policies created today begin to restrict resources.

We cannot afford to go catatonic at these threats. Individual fire agencies, professional fire organizations, and fire service leaders need to develop strategies of adaptation if things are going to work in our favor. Those strategies must be aimed at helping the fire serives to become the fittest. If we do not know what it is that we do, and if we do not know what it is that we do better than someone else, we are operating from a position of vulnerability.

Therefore, one of the best strategies for fire chiefs is to take a close look at themselves. Candidly, this can sometimes be painful. Over the last decades, we have incorporated a large number of things into the fire service, especially in the areas of emergency medical services and hazardous materials, that increase our contention that we are among the fittest, but we have not done an effective job of measuring our services, nor have we incorporated new images of ourselves into the overall organizational image of public service.

Let us look at the areas of emergency medical services and hazardous materials. These are two areas in which the fire service has been acquiring new responsibilities. In many cases, however, the roles are accepted reluctantly, as if this adaptation is an ugly growth that we would remove given the opportunity. In one series of oral boards for fire chief, I was startled to hear two chief officer candidates openly state a point of view that is contradictory to the expansion of the emergency medical services function in the fire service. One candidate said that the best thing we could do to improve the fire service would be to get back to the basics of fighting fires and elminate emergency medical services. Another candidate stated, equally unequivocally, that the only thing saving the fire service is that we are in the emergency medical services business and that the number of fires is not even an issue in his organization any longer.

We are talking about a major contradiction here. Who is adapting, and who is not? Perhaps you are not the least bit concerned about whether the entire fire service will survive. Perhaps you are more concerned about the survival of your own fire department. Those mythological budget cuts that are talked about at the national level may not mean anything to you directly, even when they shut down federal fire programs like the National Fire Academy. However, when the city manager or the board of supervisors takes a ten or fifteen percent chunk out of your budget, that will be real.

Let us focus on an adaptation strategy at the level where it means something to you. What can you do? What should you do? How should you do it? The phrase I use with regard to this area is "hardening of an organization." As mentioned earlier, whenever there is a threat to the survival of any species, only the strongest survive. That means that the survivors must be in good physical condition—lean and mean. It means that whenever you face these kinds of attacks, it is time to look into the organization and place a premium on those individuals and programs with high-performance profiles. They are your gene pool for survival.

The next strategy is to focus on opportunity instead of anxiety. If another agency within your jurisdiction is forced to take cutbacks and you have an opportunity to identify some of its functions that are parallel to those of your own organization, perhaps it is time to adopt them. Granted, this can be difficult. You do not want to go so far afield that you begin to adopt programs merely as a means of accruing favor. On the other hand, the fire service has one of the largest resource pools in its staffing levels of any city organization other than the police department. If we wish to maintain that ratio, we may have to find ways of adopting into our organization functions that will allow us to continue to be useful and to remain available.

The third strategy is to become symbiotic. Some organisms have weathered difficult times by learning to coexist with other organisms. One of the best strategies to ensure your survival is to find others who are equally concerned about survival and then form relationships with them based on your mutual interests. This can include establishing working relationships across geographical lines, functional lines, or on any other basis that will allow the formation of a consortium or alliance. This strategy is based upon establishing trusting relationships. The symbiotic relationship must be equal and balanced. One member of it does not survives at the expense of the others; rather, the survival of one member ensures the survival of the others.

There is no honor in being the last of a dying breed. I have reserved this next strategy until last because it is the one that will ensure that we are not a dying breed. In times of cutback and attack on the integrity of an organization, the one thing you should not do is eliminate training opportunities for your younger officers. Note that I say the *younger* officers. They are the future. No matter what kind of effort is required, it is extremely critical that you nurture the growth of midmanagers so that when they emerge from a threatening environment, they will have all the tools to prepare them for their newfound goal of being the new leaders. If you cannot achieve any of the previous strategies, this one is absolutely essential.

The surest way to devastate the fire service's potential future is to create one or two generations of people who cannot remember the basic values of fire protection. I have attended many meetings where it has been stated that our current officers may already have lost the moral commitment to the fire service that officers once had. The level of commitment from the next few generations will be either our future strength or our future weakness. We are in the position today to continue to cultivate generations with the basic values that gave us leadership roles in reducing the loss of life and property. We need to nurture those values so that they will still be a central theme in the profession of the fire service fifty years from today.

FADS, FANCIES, TRENDS, AND PATTERNS

One day a fire captain was telling a joke to two young firefighters. "What's a navel destroyer?" he asked. Getting no response, he gave the answer: "A hoola hoop with a nail in it." The firefighters simply stared at the captain instead of laughing. They did not know what a hoola hoop was; they had never heard of one. The hoola hoop was a fad, a temporary form of entertainment that made a fortune for someone and then disappeared from sight.

As a fire chief, you are often confronted with new ideas and recommendations for change. The fire chief has to discriminate between those ideas that are mere fads and fancies and those that are true trends and patterns that will shape the future of the fire service. Fire chiefs cannot limit their look at the future to a once-a-year budget review. As the head of the organization's management team, you must be proactive in creating the future of the fire organization.

How do you distinguish a fad from a trend? How can you discriminate between something that is temporarily acceptable and something that is becoming a permanent fixture? What tools does the fire chief have to analyze the differences between temporary and permanent change?

Let us look closely at two distinctly different terms. One is *personal preference,* and the other is *statistical significance. Personal preference* is the level of acceptability of change. *Statistical significance* is a measurement of the probability that the change will occur.

I once heard someone say that innovation passes through three stages on the way to acceptance:

Stage 1: "That won't work."
Stage 2: "It works, but I don't like it."
Stage 3: "How did we ever get along without it?"

What this tells us is that fads and fancies may become trends and patterns over a period of time. Those fads that prove to be unacceptable will eventually disappear from the scene. Those that gain acceptance do so over time.

A degree of uncertainty usually occurs when dealing with a new product or concept, but over time, this changes. Everyone has a level of personal acceptance of change. The period of time that it takes for a person to go from a high degree of uncertainty to comfort with the acquisition of the new knowledge and information that will bring about the innovation is a variable. Once a person reaches his "comfort zone," the innovation becomes acceptable.

A lot of fads and fancies are kicking around out there. A good way to locate them is to go to a fire service conference or management seminar and listen to the instructor talk about the latest buzzword or new technology. If you were charting an audience's level of acceptance, it is probable that the level would be different for each member of the audience. An instructor standing in front of the classroom espousing a trend can often gauge by body language, facial expressions, and, in some cases, verbalization just how acceptable the new idea is to the members of the audience.

We in the fire service are in a high-risk occupation. Therefore, our level of acceptance is directly related to the level of damage that can occur if something new does not work. The fire service is often painted as "conservative" and "tradition-bound" when, in fact, we are just careful. Our conservatism is based on consequence, not on philosophy.

Instead of disregarding fads and fancies, however, fire chiefs should be extremely curious about them. Many years ago there was a great controversy over the color of fire apparatus. Should they be red or lime-yellow? At that time, lime-yellow was faddish. Today we have a mixed bag of fire apparatus colors in our fire stations. We are nowhere near one hundred percent acceptance of lime-yellow fire apparatus.

Compare that experience with the introduction of the internal combustion engine at the turn of the century. Controversy also raged on that issue. It is highly unlikely, however, that you could go to any fire station in the country and see a 1,000-gpm pumper being pulled by horses. A fad becomes a trend when it becomes statistically significant. Fad-watching consists of monitoring those products and ideas that are recommended or proposed by people in isolated sets of circumstances. Almost all good ideas begin with a very small support group in some local area.

In its societal role, the fire service is impacted by fads such as hairstyles, clothing requirements, and social and moral values. In its professional role, the fire service is impacted by fads such as innovative fire apparatus design, new programs, new management techniques, and innovations in tools and equipment. A fad becomes distinctive when the level of acceptance is such that the activity or idea becomes institutionalized across geographic or demographic boundaries. For example, one might say that the concept of public fire safety education was once a fad in some areas. Today, public education programs are found in a vast array of fire service organizations. Both small volunteer fire departments and major metropolitan fire departments have public education programs. Departments in Florida, Maine, California, and Washington have public education programs. Public fire safety education is no longer a fad.

John Naisbitt, author of *Megatrends,* has made a small fortune by monitoring trends. The fire chief should become a trend-watcher also. One clue to the acceptability of a trend is the amount of space devoted to the topic in professional fire service publications. Fads are news items. Trends are the subject of journal articles and books. This type of content analysis requires that you maintain a reading list of publications that cross jurisdictional and organizational boundaries. The continued inclusion of articles that address a topic is a direct indication of the strength of the trend and of its level of acceptance in the fire service.

A classic example of the change process is the development and adoption of the Incident Command System (ICS). Fire chiefs have been "commanding incidents" for hundreds of years, but only recently have we had available a formalized incident command system. The ICS started as a tool for dealing with one specific type of problem, namely, the wildland fire. It continued to grow and evolve until today the ICS is touted as an effective interagency and cross-jurisdictional management methodology.

Things are going to change. Some changes will be temporary, and some will be permanent. One of your most difficult tasks as a fire manager is to sort recommended changes into their appropriate category—fad and fancy, or trend and pattern. The more often you accept change because it "feels good" or is your personal preference, the more likely it is that the change will be temporary. The more often you accept change because it has statistical significance and has been found acceptable by your peers and front-runners in organizational development, the more likely it is that the change will be permanent.

The only true test of successful change is time. As members of an occupation that takes great pride in its traditions, we should constantly remind ourselves that today's tradition was yesterday's controversy.

DOUBLE JEOPARDY

Many television game shows are no-brainers; they consist of a great deal of dumb luck, dumb questions, and dumb consequences. One of the exceptions to that rule is the television show called *Jeopardy!*[2] Alex Trebek, the host of the show, is an intelligent, articulate individual. The questions that are asked require a comprehensive and in-depth knowledge of a wide variety of subjects. It is no wonder that it is one of the most popular game shows on television.

---

[2] *Jeopardy!,* Columbia Tristar Television.

If you watch the show, you may notice that the person who does the best in the first part of the show is not always the person who wins in the second part of the show, "Double Jeopardy." The reason is that the first set of questions is simpler than the second set. The person who does well on the first set may or may not have the knowledge for winning the second round. At the end of the show, during "Final Jeopardy," contestants bet everything on one answer. "Final Jeopardy" consists of one single question, and the contestants risk it all right there. A person can be running in third place during the earlier phases and still win in "Final Jeopardy."

As a fire chief, you are often confronted with an increasingly complicated set of problems. Facing and solving some of the simpler ones does not always provide you with the answer when you get to a critical question that puts it all on the line. Your day-to-day decision making can often seem boring; after a while, you may even take it for granted. Periodically, you must address more complicated questions; a crisis occurs, and you have to rise to the occasion to cope with the problem. But then there is the critical decision point in which there is only one issue in front of you, and it is in the form of a career-threatening event—it may involve a budget or a decision to move forward on a major program, or it may be the juncture of other decisions that are forcing the fire service to make a major change. Sometimes we take a lot of pride in our ability to win on the simple level, but we are terrified of losing when we reach our Final Jeopardy. How do we become better prepared to cope with critical issues?

I began contemplating this particular topic when events occurred that forced fire departments to confront a series of major policy decisions. Unfortunately, the fire departments were the losers. These were fire agencies that did fairly well on day-to-day business, but when they were put to an ultimate test of convincing a political body to take a particular course of action, their recommendations went unheeded. These events involved everything from the passing of local ordinances involving residential sprinklers and questions of public safety to the dissolution of fire chiefs' jobs and fire departments through mergers, consolidations, outsourcing, and downsizing. These activities were inconsistent with good fire protection principles and created confrontation with taxpayer groups. In some cases, the fire chief and the fire department became a target.

When reviewing my notes on this topic, I noticed that this trend is parallel to the problem of inadequate financial resources. Trying to continue to provide government service levels that are consistent with past practices in restricted financial situations is problematic. The jargon of buzzwords now includes terms such as *downsizing* and *right-sizing*, and in the case of the fire service, perhaps we are now *capsizing*. I have three basic questions for your consideration:

1. Who is truly the expert on fire protection in your community?
2. Whose responsibility is it to set the level of service provided in your community?
3. Who is responsible for paying for the service levels that have been established?

At the outset, these questions look fairly simple. For example, many of us believe that the fire chief is the expert on fire protection in the community. That is why we were hired. That belief is not shared by everyone, however, judging from the clipping files that I have accumulated on this particular topic. On the contrary, it appears that expertise on good fire protection ranges all the way from retired fire department members in neighboring communities through business owners and industrialists who have some strongly worded concepts about what constitutes adequate fire protection. It is very dangerous for fire chiefs to assume that their expertise in fire fighting translates into credibility in the community in terms of being the expert on fire protection. I am not suggesting that you are not the expert; what I am suggesting is that some people do not accept the fire chief's word as the final word on the subject.

One of the reasons for this is that the fire service, in spite of its recognition of the endeavor as a fire science, seems to be deathly afraid of quantifying and qualifying its efforts. We are basically a process-oriented industry. Fire chiefs often point to a standard operating procedure and insist that it is the way something must be done. Some are unwilling to do the homework that provides the background about *why* that procedure is the most valuable, most beneficial, and most appropriate for a particular community. By relying on process instead of background information, the procedure becomes a point of contention.

If you compare this situation to the game of Jeopardy, the first set of inquiries a fire chief faces consists of the simple questions. Anybody can answer them. They become critical, however, when individuals begin to question the validity of our answers if we are unable to produce the background and facts. Not being able to tell others why we must follow a particular procedure puts us into jeopardy.

In regard to the question about who is responsible for setting the level of service in the communication, once again, I would anticipate that most fire chiefs would think that it is their job. In reality, however, most of the time, the actual recommendations that are placed on the agenda for a board

of supervisors or a city council have been carefully screened by an interme-
diary such as a city manager or chief executive officer. In essence, in many
cases, the recommendations we get to make are the ones we are allowed to
make.

The types of scrutiny that are applied to our written recommendations
as they proceed through the approval process are not unlike those of our
"community experts" who frequently disagree with how we believe we need
to do things. The difference is that it is a closed shop. The individuals who
get to scrutinize our recommendations are often our fellow department
heads, and specifically the finance director, personnel officer, and chief
executive. This is our version of Double Jeopardy. The degree to which the
recommendations are unmodified as they proceed through the process
speaks volumes about the trust and credibility of the fire chiefs. When an
attempt is made to modify these recommendations, usually through the use
of fiscal arguments, we often face difficulties that move us into in our Final
Jeopardy phase.

At this stage of the game, it is not so much that people want to disagree
with us in setting our level of service as it is the fact that we are in competition with other services. The more that we are able to qualify and quantify
what we are doing, the better off we are. Simply pointing to our manual of
operations and saying that this is the way it has always been done is rapidly
losing its credibility as an input to the decision-making process. For example, I participated in one discussion involving a policy issue of a fire agency
in which the chief executive officer in the community challenged the chief
fire officers to predict the consequences of failing to follow his recommendation. The response included a lot of mealy-mouthed, fuzzy-edged words
like *maybe, could,* and *might.* The chief executive was relentless and refused
to let the chief officers off the hook until they made a prediction that a certain event would transpire.

Most city managers do not want to intimidate their fire chiefs, but they
do want us to have solid answers that are defensible when it comes time to
face public scrutiny of our recommendations. We do not actually set the service level in our communities; it is established by the budgets and public
policies that are adopted by the authorities that we serve. Public scrutiny of
our recommendations determines the level of service that will be provided.
In the majority voting process, the perception of a single individual regarding whether he supports the level of service being recommended can determine the ultimate level of service.

There are probably many of us in the fire service who can recall the
days when our budgets were basically made up of the previous year's

expenditures along with an inflation factor. We used to argue about who got the inflation increase, but seldom did we worry about the core funding component. Now we talk about "structural imbalance in the fiscal area," which means that the taxing system in many of our communities is inconsistent with the service level that has been provided in the past. It is virtually impossible for us to continue to grow without reprioritizing the governmental services that are to be provided. This reprioritization process is painful and, at the same time, virtually impossible to avoid in any community that has a structural imbalance.

All three of the areas raised in the preceding questions and discussion provide us with a snapshot of the future. There probably will be an increasing number of individuals in our communities who will compete with the fire chief as being the expert on what constitutes adequate fire protection. In addition, more and more people are likely to exercise leverage regarding our recommendations on levels of fire protection service as different disciplines compete with us for the existing resources. Finally, there probably will be more and more public scrutiny over the final adopted budgets of fire protection agencies with a continuing emphasis on the value for the dollar being spent.

As a fire chief, what strategy should you engage? There are a wide variety of potential tactics and strategies that can be used with the exception of one. The one that probably will not work is to stick with current practices. And that holds true even for departments that are doing very well. The price we must pay for keeping fire protection a priority in our communities is eternal vigilance. There is no such thing as a status quo. Any chief officer who practices this form of organizational myopia will be blindsided sooner or later. It will happen to some more quickly than others; some might take a great deal longer. But the status quo is simply not a viable alternative.

You can go to almost any bookstore these days and find numerous books on management and leadership. Everybody has a theory. I am not going to promote any one theory here; rather, I will make one observation about the role of the fire chief in being responsible for the destiny of a fire protection agency. It can be stated simply as, "If you don't have a plan, plan on losing the game." Budgets are not a once-a-year question any longer. You must have the ability to work through the system and deal with your customers and constituencies on day-to-day events in order to lead the community into fully accepting and supporting your fire agency's activities.

This kind of game plan requires a great deal of innovation and entrepreneurship. The essential component is the application of energy by the chief officer. In the future, behaviors of apathy, lethargy, and mediocrity in fire chiefs will result in drastic deterioration in organizations. The application of energy literally means paying attention all the time, every day, in every way to what is going on in the community so that you can be a part of the community's, not just the fire agency's, decision-making process.

As you can see, the fire chief's job is not getting any easier. If you reflect on the history of the fire service, however, you will realize that it has never been an easy job. The only thing that has changed from the past to the present is that the questions to which we must respond are becoming more difficult all the time.

On the game show *Jeopardy!*, when Alex Trebek calls for the final question, there is always the possibility that all three people will have the right answer. However, only one is allowed to win. Two things determine who becomes the winner. The first is how much the person has in his bank account when he reaches "Final Jeopardy." The second is how much the person is willing to bet on the answer to the final question. Sometimes a person with a lesser amount will win because he knows the answer and bets everything. Other times a person who has the wrong answer will still win because he has so much money in the bank and bets so little of it. And so it is with your Final Jeopardy. The investment in your community and the amount of time you spend getting ready for critical decision making is much more important than a single event itself.

Sometimes we use short, succinct phrases to express things that are much more complex and sophisticated. For example, the hippies used to say, "Be cool." Trying to be cool is really a very complex task. Another popular phrase was "Looking good," which meant that a person "had his act together." We in the fire service need to manifest these three clichés at the time of an emergency. In other words, each and every fire officer who finds himself dealing with an unfolding emergency has to be cool and look good while getting his act together. We have a term for this, although you will seldom find it in the definitions and acronyms that have been offered as a means of controlling the thought processes during an emergency. The term is *command presence.*

Command presence is simple to describe but extremely difficult to achieve. Command presence is an outward exhibition of total control of yourself in a situation that is being recognized and respected by those who are observing your behavior at the time of an emergency. In actuality, it is probably easier to identify the lack of command presence than to identify its existence. You probably have seen fire officers exhibit aberrations of their own personalities under pressure. Some individuals create stress in others with this behavior. You would, no doubt, recognize some of the symptoms: talking loud or shouting orders, the use of profanity to place emphasis upon a certain task, and dysfunctional behavior, such as running back and forth without actually achieving any given task. The lack of command presence is readily identifiable when a person in charge of a fire incident loses control of himself. Lacking control over themselves, such individuals usually do not have much control over the emergency at hand.

If you say that this has never happened to you, then you must be a rarity in this business. Very few individuals who achieve the rank of fire officer go directly to the stage of personal development where command presence is always there. This is not to say that people *always* panic under pressure, but sometimes there is a feeling of anxiety that arrives at the same time that the flush of adrenalin to your cardiovascular system begins to accelerate all of your other senses. Even the most seasoned individuals have experienced it at one time or another in their careers. If they are realistic, they will admit it.

How do you achieve command presence? Once again, it is easier said than done. But we can look outside the fire service and find examples of how command presence is built into individuals with tremendous responsibility at all times. The classic example is found in the airline industry. If you listen to a tape recording of an aircraft in distress, it is almost eerie to hear the dialogue between the pilot and ground control. Despite the fact that the aircraft is suffering from severe damage or, in some cases, is even out of control, the pilot maintains control over himself, clearly describing the conditions to which he is being exposed and the course of action that he is following to remedy the situation. Airline pilots are the deans of the "cool school."

There is a reason why airline pilots maintain command presence. The reason is so simple that it often defies observation. The pilot realizes, through training, education, and previous experience, that the only way the pilot and the rest of the passengers on the aircraft are ever going to survive is if control is maintained over the situation. This discipline has been bred into pilots practically from the first day that they sat in the pilot's seat. The operative word here is *discipline*. The word *discipline* does not in any way imply punishment. In fact, *discipline* comes from the word *disciple,* which means student. The reason that pilots are disciplined individuals at the time of an emergency is that they have been students of those conditions for so long that they do not feel they are part of the problem, but rather, they feel that the only solution to the problem at that time is in their hands.

The same thing applies to the fire service and to your role as fire chief. An emergency is something that is happening to another individual or group; it is a condition that is going to continue to deteriorate unless someone takes total control and provides a remedy to the set of circumstances that is creating the emergency. This is particularly unique in the fire service because we are one of the few public safety agencies that responds to emergencies that are still in progress.

Many public safety services respond to emergencies that are, for the most part, already over and merely require a report to be filed. For example, police officers often respond to situations that have already unfolded; in many cases, the perpetrator has left the scene, and law enforcement personnel are primarily responsible for the collection of data. When a police officer does get involved in an emergency that is life-threatening, such as an exchange of gunfire or a high-speed pursuit, it is dramatic and highly visible, but such emergencies are actually a minority of law enforcement activity. Fire service personnel on the other hand, respond to a significant portion of emergencies in which their arrival time is critical to the ultimate outcome in terms of destruction of life and property.

This phenomenon contributes to some of the problems associated with the maintenance of command presence on the part of fire officers. In the vast majority of fire service emergencies, the resources that are immediately under our control, that is, an engine company, are adequate to deal with the emergency circumstances in a relatively short time frame. This is a time frame of less than seven to ten minutes, and probably ninety percent of the emergencies we respond to fall into this category. We rapidly grow accustomed to a decision-making process where we can identify an outcome in a relatively short period of time and have expectations that will be achieved in this short time frame.

On the other hand, when we face emergencies that are beyond the resources that are immediately available, we have a tendency intellectually to try to compress the situation. When a situation is beyond the realm of control in a short period of time, our minds tend to speed up the process in hopes of compressing the control element back into a time reference that is acceptable. However, that simply is not going to happen under a high-stress set of circumstances. When an emergency exceeds all of your resources or, for that matter, places you in a position of making short-term critical decisions with long-term impacts, you have to be prepared. Your mind has to be ready for that eventuality. It has to be disciplined.

Early in my career, I was taught a valuable lesson along these lines. I was attending a school in which there was a senior fire officer from a neighboring city. I was a young fire officer with a limited amount of experience with fully involved structures. One day when he was paged to respond to a structure fire in the neighborhood, I rode along with him to see how he dealt with the issue.

We saw the header from far away. Although I had no responsibility for the incident myself, I readily admit that my adrenalin started pumping, my heart rate probably went up, and all of the "fight instincts" that have been with humans since the cave-dwelling days started to come into play. When I looked at the chief officer, he appeared as calm as though he were driving to pick up his lunch.

Upon our arrival at the fire scene, I disembarked the vehicle and made four or five rather hasty laps around the building to see what I could find out about the situation. As I looked back, I noticed that the officer was very much in control of himself. He was standing alongside his vehicle, studying the fire problem. I could not measure his powers of observation at that point, but I noticed that he appeared to be rock steady in his position. Slowly, he put on his turnout coat, placed the helmet upon his head, and began to talk very calmly and rationally into the radio as he directed the incoming apparatus.

After the fire was over, I asked him, "How do you stay so cool when all of that is going on?" His answer was classic. "Well, Ron," he said, "you have got to remember that I didn't start that fire. My job is to make sure that it goes out."

And that is your job. Your responsibility is to make sure that emergencies are handled in the most effective manner possible. It is vital to the professional development of the fire service that individuals who are held accountable to be in charge of incidents look upon themselves as the "solution." Command presence is every bit as much of the inventory of a good fire officer as is apparatus, manpower, and all of the programs in which he engages prior to the ringing of the bell that calls him to respond to an emergency. Moreover, we must be able to demonstrate command presence in serious situations that are nonemergencies, but which nevertheless demand calmness and discipline.

Some individuals believe that command presence comes with the job. Pinning a badge on a person's chest and calling him battalion or fire chief, however, does not result in the proper exhibition of that kind of behavior. There are some specific techniques in which fire officers should engage if they want to assure themselves that they will have command presence.

First and foremost is training and education prior to assuming the role. It is axiomatic that an individual feels more in control of a set of circumstances when he is totally knowledgeable regarding the short- and long-term impacts of every decision that he makes. The more a person knows about important factors such as building construction, fire chemistry, management methodology, and so on, the more likely he is to maintain control. It is just as important for a fire officer to understand the anatomy of the decision process as it is for a doctor to understand the anatomy of a human being.

The second aspect of command presence is to maintain yourself in good physical condition. Like a well-trained athlete, an individual will perform a lot better under stress when his body is capable of absorbing the physical aspects of stress with a minimum amount of impact. Granted, the function of decision making is mostly a cerebral exercise; however, if a person is not in good physical condition and is easily debilitated, the mental processes will begin to deteriorate. There is even a term for that; neurosthenia. This is where an individual who has been exposed to stress for a long period of time enters into a fatigue factor and starts making decisions that are improper and inaccurate.

Another technique for improving your ability to exhibit command presence is to study individuals who already have it. To paraphrase an old get-rich-quick scheme, if you want to know how to be in control, study someone who is effective in maintaining control. In a sense, this is another use of the concept of role modeling. Patterning your behavior after another person's behavior is not being a copycat. Instead, it is a means of short-cutting the time it takes to achieve your own personal style.

Lastly, one of the best techniques for improving command presence is to engage in what I refer to as the Walter Mitty approach.[3] The approach is to play mental games with yourself in terms of how you would like to handle decision-making processes and to daydream your way through certain kinds of scenarios. Walter Mitty's life, as some of you will recall from reading the short story, was a series of fantasies, one after the other. I am not suggesting that fire chiefs live in a fantasy world, but visualization is one of the best ways of developing a mental frame of reference. It is a form of dress rehearsal in your mind. This technique, far from being unsophisticated, is actually an important part of the training process of Olympic and professional athletes.

In summary, command presence is very difficult to achieve. You cannot take a class and, upon graduation, be annointed as a person who possesses command presence. Instead, it is an outcome of many other factors in the professional development of a fire officer. The lack of command presence erodes the credibility of the fire service. The fire chief who has command presence helps to establish confidence and credibility on the part of the profession.

SELECTED READER ACTIVITIES

1. Review your own list of "things to do" for the department and prioritize the items on that list.
2. Describe the latest fads and trends that are affecting the fire service. Which ones will survive? Which ones will not last?

---

[3] Ron Coleman, "Walter Mitty Firefighting," *Fire Chief* (April 1984).

3. Go back to the day you entered the fire service. Describe the changes that have occurred since that time. Identify those changes that have had positive consequences.

4. Now that you have undergone the basics of transition, make a list of the things for which you were prepared and a list of those that caught you unaware.

5. Pick a person whom you respect for his ability to remain calm under stress. Discuss with him how he keeps from losing his temper or control under pressure.

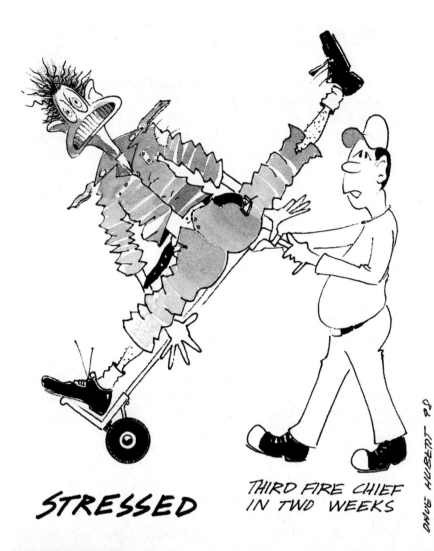

STRESSED

THIRD FIRE CHIEF
IN TWO WEEKS

DAVE HUBERT 90

CHAPTER 16

Is This a Marathon or a One Hundred Yard Dash?

Reducing Stress

The objective in this chapter is to focus on personal stress management. Once a person has gotten control of the events described in the previous chapters, the sheer momentum of the organization will create a need for the chief to find ways of bleeding off pressure. The techniques discussed in this chapter include an emphasis on both physical and mental health.

Membership requirements to join the Burnout Club:
1. *Take everything personally.*
2. *Don't ever delegate; do it all yourself.*
3. *There are twenty-four hours in a day. Work all of them except when you are asleep, especially holidays.*
4. *Never say no, and when you do, feel as guilty as possible about it.*
5. *Assume responsibility for everyone else's problems.*
6. *Be as rigid and inflexible as possible and a perfectionist to boot.*

—Anonymous

STRESS AND THE FIRE CHIEF

It is very difficult anymore to pick up a fire magazine without reading some kind of comment about postincident stress. A lot of time and attention has been focused on the recovery of personnel who have faced a traumatic set of circumstances in a short period of time, such as in a major airplane crash, major trauma incident, or loss of a fellow firefighter. This work is long overdue. Most of us recognize it when we are under stress. The pain and discomfort that comes from this kind of stress are sometimes independent of the cause of the stress. There are very strenuous sets of circumstances that create in a person a feeling of exhilaration instead of stress, but there are other sets of circumstances when limited action or frustration can result in a high degree of stress on the human body.

Truly, the most common examples of stress-inducing circumstances have nothing to do with emergencies. As a fire chief, you will often face stress that has absolutely nothing to do with a field emergency. Instead, it will be a crisis or condition having to do with the management and administration of your fire department. This type of stress, while perhaps not as glamorous or topical as postincident stress, is responsible for a lot of casualties at the top level of fire departments.

Eliminating stress from behind the chief's desk is not easy. If stress is allowed to become persistent or recurring during an individual's career, the job of running a fire department often can be more stressful than that of running emergencies. For example, how much stress do you think is induced in a chief officer when he is informed by the city manager or budget officer that there must be a ten percent reduction in the overall fire department budget? What happens to a chief's blood pressure and heart rate when a council member announces, in an open forum, that he would like to see the fire department become consolidated with a neighboring agency, become privatized, move to a public safety concept, or reduce the staffing from the four-person to three-person engine companies for economic reasons? Yes, there is just about as much stress behind the desk as there is on the end of a nozzle, and it is not nearly as much fun. You had better start planning for ways in which to deal with stress during your original transition, because if you do not get off on the right foot, you will never get there.

Some individuals believe that stress simply "goes with the territory" and expect fire chiefs and chief officers to handle it in silence in spite of the fact that it may be as physically and mentally debilitating as some of the stress reactions that occur after major emergencies. Stress that occurs in emergencies is a serious problem that must be dealt with both now and in the future. Postincident stress is becoming more of a problem now that we have large numbers of individuals serving as emergency responders who have not faced these kinds of conditions in the military, and there are increasing numbers of emergencies that are so catastrophic in nature that we, as a profession, have never faced anything like them before. My emphasis on stress behind the desk does not contradict the problem of stress in emergencies; rather, such stress is a parallel problem that needs to be recognized by those who are in key staff officer positions or who aspire to the positions. Fortunately, there are many techniques that individuals can utilize to reduce the amount of stress on themselves and the organization.

What can you do to keep stress to an absolute minimum? You will never get much information on this out of the contemporary textbooks on the fire service because it is not considered to be an element of fire management or technology. Yet, a great deal of time and effort have been spent on stress in other fields of endeavor. The late Hans Selye wrote on the topic many times. In reviewing some of Selye's work, it became clear to me that one of the biggest causes of stress for individuals who are functioning in top management positions is ambiguity—not being able to control a set of circumstances results in a psychiatric implication. Not being able to adequately deal with a set of circumstances also results in frustration, which often creates stress as well.

Therefore, it is axiomatic that the more you remove ambiguity from the job environment, the lower your stress level may become. Note that I use the word *may*. There are no guarantees in the development of a methodology to reduce stress because each technique has degrees of effectiveness that are directly related to the amount of emphasis by the user.

Some of the techniques are based on rules that may sound clichéd or trivial, but they are not. To the contrary, in talking to hundreds of chief officers, I have found that sometimes our problems are caused by a failure to follow basic principles. We might be very sophisticated in the development of some aspects of fire protection technology, but we have been very circumspect when it comes to dealing with some basic management principles.

One of the first rules for removing ambiguity in the job is to know your complete job description and what is expected of you. This was discussed in the first section of the book. In talking with people who were in conflict with their superiors, I have been amazed to discover a major difference of opinion between the individuals and their bosses regarding the individuals' job expectations. Make sure this never happens to you. It can happen in a variety of circumstances if you do not perform a reality check once in a while. It often occurs when an individual who has been the second officer in an organization is promoted to chief upon the retirement of his superior.

The number two position in an organization is often radically different from that of the chief, and yet the companionship associated with the two jobs may result in the number two person believing that he is eminently qualified and prepared to pick up the reins. Depending upon the thoroughness of the superior officer in briefing the subordinate, a lot of role ambiguity can occur when the promoted person assumes the new position. In one case, a fire department promoted an individual who had been extremely involved in the "nuts and bolts" of the organization. Upon his promotion to fire chief, he found himself becoming more and more bogged down by dealing with all of the details, and he got into conflict with the city manager because he failed to meet major deadlines on projects such as budgets and intra-agency activities in the community. The harder this person worked, the more he fell behind. He never utilized effective delegation techniques because he believed that he was doing the job he was supposed to be doing. The city manager, on the other hand, recognized the individual's talent for technical functions but wanted the individual to function on a different plane. It was only a matter of time before stress drove him off the job.

One of the most *stress-reducing* relationships that a fire chief can cultivate is with his superior. Develop a thorough understanding between yourself and your superior regarding what is expected of you. Discuss expectations on a daily basis, weekly basis, monthly basis, annually, and overall professionally. Good honest communication between you and your superior regarding expectations is one of the first steps in removing conflict over performance. This activity lowers stress by its very nature.

Another inherent danger in the promotional situation is the assumption that past relationships are always one hundred percent reliable for making future decisions. Two conditions can change in any fire chief's situation that may alter job expectations. The first is the change of a superior, for example, city manager or other senior executive officer, and the second is a change in the policy direction of the organization, for example, a major change in council policy or direction. In both cases, previous expectations may have to be reviewed. I am not suggesting that the fire chief become a chameleon who constantly attempts to rise to other people's expectations; to the contrary, being honest with oneself and being able to relate one's own expectations to one's superior is an important part of removing ambiguity. If an individual is intimidated by his superior and fails to ask the right questions regarding these relationships, minor problems may be blown out of proportion.

Rule number two is simple. Set realistic goals and then do everything you can to achieve them with the resources at your disposal. Setting goals that are too long-term or goals that are extremely altruistic or unrealistic will undoubtedly result in frustration and stress. Making a realistic appraisal of the resources and capabilities that are available to you is equally important.

One chief with whom I discussed this issue took over a fire department that had rather complex fire prevention problems. The fire prevention bureau was ill prepared to rise to the occasion of resolving those problems in a reasonable time frame. Instead of asking staff to put in more hours or to "get tougher" on code enforcement, this person took a different tack. He set a goal for himself of upgrading the technical competency of the fire inspectors and devoted the first few years of his position to training the fire prevention bureau to undertake the task of code enforcement.

One has to recognize, in an organizational context, that setting realistic goals is part of the self-image that is created for persons in the organization. Unrealistic goals often give people a sense of failure even when they are actually achieving a great deal. It is important to allow goals to escalate as competency and capability enter an organization, instead of imposing goals on an organization in an attempt to achieve herculean accomplishments. A good analogy in sports is the breaking of the four-minute mile. The four-minute mile was supposedly an impenetrable barrier, and it did take a great deal of time before Roger Bannister broke it. Once it was broken by Bannister, however, several individuals were able to run a four-minute mile in rapid succession.

Goal setting and the measurement of objectives in an organization are a part of stress reduction. They establish a relationship between task activity and achievement that allows persons who use the system to monitor their own activity levels.

Stress-reduction technique number three is also very basic. Have a game plan and work it. Develop a particular approach to your job that you can feel good about, and become successful in utilizing it. It is a good idea, for example, to determine what part of the day you can be the most productive, creative, or efficient in getting your job done and then, in order to remove ambiguity from the situation, devote that time to accomplishing things that have to be done and leave the remainder of your time open for the random activities that sometimes impact on the fire chief's job. Do not be reluctant to set aside this time and close your door to enforce it. Sometimes subordinates feel that a closed door means a closed mind, but when a closed door means that a person is devoting a specific block of time to accomplishing a specific objective, subordinates usually honor the sanctity of that type of decision. And, as we mentioned in earlier chapters, you cannot forget the impact of your stress on spouses and children. Overcommitment is not just stressful on you; it is stressful on others as well.

Stress reduction, when it comes to the game plan, has a lot to do with how many different plays you have in your game plan. As previously mentioned, I utilize a lot of the time in my vehicle to dictate memoranda or letters. Some chiefs do not like to use dictation equipment and feel uncomfortable talking to a machine. That is fine. What is important is not the specific technique you use, but whether you *have* a specific technique to use. Those chiefs who do not like to use dictation equipment often survive quite nicely by working out a system of devoting time to handwriting memoranda with a minimum of interference.

Among the specific plays about which you might want to be concerned is how to handle deadlines. Some people do not mind being under the stress of having to complete something by a time certain, while others are deeply disturbed by it. The individual who decides what his level of ambiguity is in regard to deadlines will have less stress. For example, I do not mind working under deadlines because the deadlines intensify my concentration on specific topics. On the other hand, if you are stressed by deadlines, it is extremely important that you have a game plan for finishing projects long before their deadlines. Give yourself a break by not creating the crisis.

All of us have seen various techniques promoted to help us prioritize our daily activities and our "to do" lists. Books on time management are extremely popular. We spend a lot of time talking about time management. The one thing that you need to remember about it is that you are not really managing time, you are managing tasks, projects, programs, and activities that must be completed within a certain time frame. The system cannot be more important than the accomplishment.

One of the things that seems to reduce stress a great deal is to relate reward to accomplishment. It is not enough simply to prioritize all your tasks and drive yourself from dawn to dusk in an attempt to deal with all the "Type A" items on the list. It is more satisfying to accomplish specific tasks and then reward yourself with the psychological freedom to wander from the task for a short time before focusing on another. This can take the form of a coffee break, a brief conversation with a coworker, a walk in the park, or any other psychological reward. The important thing is that prioritizing tasks has to have with it, for stress reduction purposes, a reward mechanism so that each and every accomplishment gives you a feeling that not only are you achieving, but you are feeling better about it.

Changing scenery or changing your workplace is almost as important in handling prioritized tasks as the list of priorities itself. One chief officer I worked with and I had a "ritual" to deal with the closure of projects. Invariably, whenever we finished a major project, one of us would suggest that we go to a particular shop in town and indulge in a yogurt sundae. Each time we engaged in this activity, we would do something that made it unique, such as, changing the types of toppings we had or eating the food in a different place. After a while, the ritual itself became a reward. It reached a point where we often felt like we were working to finish a project so that we would have the opportunity to engage in the reward activity.

Rewards are different things to different people. Personality orientations result in people having a variety of perspectives on what constitutes compensation. What you choose as your means of reward is irrelevant. What counts is that you establish a means of rewarding yourself for accomplishing things so that, in spite of the trials and tribulations you have in finishing a particular project, the reward provides compensation for the achievement in your mind.

The next rule of stress reduction is one that Harvey Anderson, a retired Los Angeles County chief officer, constantly preached to the fire service. He advised chiefs to stay in good shape. Granted, by the time we reach the level of senior officer in the fire department, most of us are not exactly ready to compete in a decathlon. But that does not mean that we cannot take good care of ourselves and be concerned about our general health. Good stress-reduction techniques in this area include following a good diet, having some form of exercise regimen, having an annual physical examination, and avoiding excesses.

Fatigue and stress are often mistaken for one another. Stress is usually something that results in fatigue, and not the other way around. Learning to pace oneself with regard to physical ability is an important part of keeping one's fatigue level under control. It is sometimes difficult to keep from facing a fatiguing set of circumstances, such as long hours during emergencies or budget sessions, but an individual who is in reasonably good condition can often endure far longer than her contemporaries. But that does not mean that you should not take time to relax.

The next suggestion for stress reduction has to do with having a sense of humor. A lot of stress is created in the workplace by the inability to look at our situations, coworkers, subordinates, superiors, and ourselves with any degree of humor. I am not talking about making fun of people or being a full-time comedian, but fun and humor have their place in the office. Many organizations will not allow their personnel to have birthday parties or recognize specific events in the office setting for fear that it will diminish the professionalism of the organization. I do not wish to get into a debate over the amount of time that can be devoted to such things, but it is important to periodically have social interaction in an office setting for purposes of stress reduction.

Humor in the workplace does not consist of conducting practical jokes. It has to do with looking at oneself in a realistic way and recognizing one's own weaknesses and inconsistencies. It should never be at the expense of others. It is all right to laugh in the work environment, even under stressful circumstances, but we should always be laughing at ourselves.

I recall one humorous incident that resulted in considerable stress reduction in connection with a budget hearing. The city manager had told me and the other city department heads that a very critical eye would be placed on all the items in our budgets because of lack of revenue. This city manager constantly used the phrase *bare bones budget* in his presentation about the budgets we were expected to submit. A group of my fire officers who were amused by the phrase arranged to cook a set of spare ribs at the firehouse that evening. Upon completion of the meal, the ribs were washed, cleaned, and soaked in a bleach solution until they were snowy white. The budget that was presented to the city manager on the day it was due consisted of a box of bones with all of the line items of our budget very carefully printed in black ink on them. Unfortunately, this did not result in any increased revenue in the fire department's budget, but it did break the tension of the situation and definitely changed the mood of the discussion. As a result, there were some compromises made about reductions, and there was no loss of personnel in the organization.

The area of humor is difficult at times because one person's humor can be another person's sarcasm or blind spot. There is a lot of material available on humor in the workplace. Perhaps some study of this material by those of us who are operating in high-stress circumstances is in order.

My last suggestion for stress reduction is the Walter Mitty approach referred to earlier in the discussion of command presence. One of the best ways of reducing stress is to anticipate what stressors may be coming in your general direction and do everything in your power to role-play your way out of the situations. This means anticipating problems and conflicts with superiors and subordinates, anticipating what will occur at a city council meeting during a budget hearing or code adoption process, and so on. While it is unlikely that you still hit upon the exact series of events that will occur, by role-playing your way through the scenario, you may be able to predict what you are going to do. Most topflight professionals in the field of conflict resolution are adept at rehearsing events before they occur. Witness how trial attorneys prepare for courtroom presentations or how professional athletes engage in dress rehearsals before they play for a championship.

So, although a lot of stress does occur at the scene of emergencies, some of it occurs while we are sitting behind our desks or standing in front of political bodies. Stress is not going to go away, it is part of our lives. The more complex the life, the more stressful it is. You can do a lot to ensure your survival as a command officer by developing approaches that will reduce your personal stress levels.

SELECTED READER ACTIVITIES

1. Make a list of things that cause stress for you. Are there any things on that list you can do something about? Do it.

2. Review your expectations today with those you had when you took the job. What has changed? What has remained the same? Can you do anything about them? Do it.

3. Make a list of how you reward yourself. How many times have you used these rewards?

4. Answer these three questions: Are you in good health? Are you having fun? Why not?

5. Answer the following question: If you were given the opportunity to revisit your original decision to accept the chief's job, would you do it again? No matter whether the answer is yes or no, provide the reasons for your choice.

References

America Burning. The Report of the National Commission on Fire Prevention and Control. Washington, D.C.: U.S. Government Printing Office. 1973.

Bachtler, Joseph, and Thomas F. Brennan, *Fire Chief's Handbook*. 5th ed. Pennwell Publishing Co. 1995.

Blanchard, Ken. *The One-Minute Manager*. New York: Morrow. 1982.

Coleman, Ronny. "Walter Mitty Firefighting." *Fire Chief*. April, 1984.

Fink, Stephen. *Crisis Management: Planning for the Inevitable*. New York: Amacom. 1986.

Gellerman, Sol W. "Why Good Managers Make Bad Ethical Choices." *Harvard Business Review*. July/August, 1986.

Gray, John. *Men Are from Mars, Women Are from Venus*. New York: HarperCollins. 1992.

Keirsey, David W., and Marilyn Bates. *Please Understand Me: Character and Temperament Types*. Del Mar, CA: Prometheus Nemesis Book Co. 1984.

Kuhn, Thomas. *The Structure of Scientific Revolution*. Chicago: University of Chicago Press. 1996.

Lustberg, Arch. *Testifying with Impact*. Washington, D.C.: American Society of Association Executives. 1983.

MacKay, Harvey B. *How To Swim With the Sharks Without Being Eaten Alive*. New York: Morrow. 1988.

Plotnick, Arthur. *The Elements of Editing*. New York: Macmillan General Reference. 1997.

Shaw, Sir Eyre Massey. *Fire Protection, Operations, Machinery and Discipline of the London Fire Brigade*. 1886.

Shertzer, Margaret. *The Elements of Grammar*. New York: Macmillan General Reference. 1996.

Strunk, William I., and E. B. White. *The Elements of Style*. Needham Heights, MA: Allyn & Bacon. 1995.

Sun Tzu Li. *The Lost Art of War*. Translated by Thomas Cleary. San Francisco: Harper. 1997.

Sun Yat Tzu. *The Art of War*. Translated by Thomas Cleary. Boston: Shambhala Publishers. 1988.

Suggestions for Further Reading

This reading list consists of books that relate to the concepts discussed in the chapters of this book. The reader will note that some of these books are fairly old, almost "classics" in the field, while others are newly printed. This suggested reading list is being offered as a supplement to the text because a text of this size cannot contain all of the specifics that a fire chief needs to perform the job in contemporary times, and these books further define the body of knowledge that is needed by a fire chief. If you read these books, you will perform your job as fire chief better.

I have two very specific suggestions for you. First, get a library card and use it. Second, join a management book club and begin to develop your own library. The twenty to thirty dollars a month that it will cost you is not only tax-deductible, but it is an investment in yourself. I have read all of the books in this list, and I also own all of them. You do not have to own them, since most of them are readily available from a well-stocked library; however, you may want to purchase some of them to use as desk references during your career. The management and leadership sections of most libraries have more books than you will ever have time to read, but having them available to you on your own bookshelf makes them more accessible and, therefore, more useful.

Chapter 1   When Were the Good Old Days? A Historical Perspective

Coleman, Ronny J., & John A. Granito, *Managing Fire Services* (Washington, D.C.: International City and County Management Association, 1988).

Fire Chief's Handbook (New York: Pennwell Publishing, Fire Engineering, 1995).

Hoffer, Eric, *The Ordeal of Change* (New York: Harper & Row, 1963).

Chapter 2   Getting into the Playoffs: The Decision to Try for Chief

Covey, Stephen, *Principle Centered Leadership* (Summit Books, 1991).

Covey, Stephen, *The Seven Habits of Highly Effective People* (New York: Simon & Schuster, 1989).

Peters, Thomas, & Robert H. Waterman, Jr., *In Search of Excellence* (New York: Harper & Row, 1982).

Chapter 3   Mirror, Mirror, on the Wall: Meeting Expectations of the Process

Hertzberg, Frederick, *Work and the Nature of Man* (Cleveland, OH: World Publishing, 1966).

Keirsey, David W., & Marilyn Bates, *Please Understand Me: Character and Temperament Types* (Del Mar, CA: Prometheus Nemesis Book Co., 1984).

McGregor, Douglas, *The Human Side of Enterprise* (New York: McGraw-Hill, 1960).

Chapter 4   What They Never Teach You at Fire Academy: What You Have to Gain and Lose by Becoming Chief

Howard, Philip, *The Death of Common Sense* (New York: Random House, 1994).

Kouzes, James M., & Barry Z. Posner, *Credibility* (San Francisco: Jossey-Bass, 1993).

Osborne, David, & Ted Gaebler, *Reinventing Government* (Reading, MA: Addison-Wesley, 1992).

Chapter 5   Why Do You Really Want to Become Chief? Personal Commitment to the Process

Bay, Tom, & David McPherson, *CYA: Change Your Attitude* (Newport Beach, CA: MacBay Presentations, 1997).

Block, Peter, *Stewardship: Choosing Service over Self-interest* (San Francisco: Berrett-Koehler, 1993).

Jaffe, Dennis T., Cynthia D. Scott, & Glenn R. Tobe, *Rekindling Commitment* (San Francisco: Jossey-Bass, 1984).

Chapter 6   Are You Ready for the Spotlight, or Are You a Deer in the Headlights? Developing Specific Job Skills

Jacob, Ernst, *Writing at Work: Do's and Don'ts and How-to's* (NJ: Hayden Books, 1976).

Lambert, Clark, *The Business Presentations Handbook* (Englewood Cliffs, NJ: Prentice-Hall, 1988).

Leech, Thomas, *How to Prepare, Stage, and Deliver Winning Presentations* (New York: American Management Association, 1982).

Plotnick, Arthur, *The Elements of Editing* (New York: Macmillan General Reference, 1997).

Poe, Roy W., *Handbook of Business Letters* (New York: McGraw-Hill, 1983).

Shertzer, Margaret, *The Elements of Grammar* (New York: Macmillan General Reference, 1996).

Spinrad, Leonard, & Thelma Spinrad, *Speaker's Lifetime Library* (Paramus, NJ: Prentice-Hall, 1997).

Strunk, William Jr., & E. B. White, *The Elements of Style* (Needham Heights, MA: Allyn & Bacon, 1995).

Chapter 7 Putting Your Best Foot Forward: Actually Applying for the Job

Baker, Wayne E., *Networking Smart* (New York: McGraw-Hill, 1994).

Connor, Dennis, *The Art of Winning* (New York: St. Martin's Press, 1988).

Kaplan, Burton, *Everything You Need to Know to Talk Your Way to Success* (New York: Prentice-Hall, 1995).

Oldham, John M., & Lois B. Morris, *Personality Self-portrait* (New York: Bantam Books, 1990).

Chapter 8 Transition, Transplants, and Training: Moving into the Chief's Seat

Badaracco, Joseph L., Jr., *Defining Moments: When Managers Must Choose Between Right and Right* (Boston: Harvard Business School Press, 1997).

Bridges, William, *Job Shift* (Reading, MA: Addison-Wesley, 1994).

Bridges, William, *Managing Transitions* (Reading, MA: Addison-Wesley, 1991).

Sun Yat Tzu, *The Art of War,* trans. Thomas Cleary (Boston: Shambhala Publishers, 1988).

Chapter 9 Suddenly a Shot Rang Out: Preparing for Criticism from Others

Fisher, Roger, & William Ury, *Getting to Yes* (New York: Penguin Books, 1981).

Matthews, Christopher, *Hardball* (New York: Summit Books, 1988).

Solomon, Muriel, *Working with Difficult People* (Englewood Cliffs, NJ: Prentice-Hall, 1990).

Chapter 10 Cleaning Up Your Act and Clearing Out the Closet: Examining the Department

Fire and Emergency Service Self-assessment Manual (Fairfax, VA: Commission on Fire Accreditation International, current ed.).

Weaver, Richard G., & John D. Farrell, *Managers as Facilitators* (San Francisco: Berrett-Koehler, 1997).

Chapter 11 Juggling Eggs, Chain Saws, and Hand Grenades: Managing Your Time

Mayer, Jeffrey J., *Time Management for Dummies* (Foster City: IDG Books, 1995).

Mayer, Jeffrey J., *Time Management Survival Guide for Dummies* (Foster City: IDG Books, 1995).

Silver, Susan, *Organized to Be Your Best* (Los Angeles: Adams-Hall, 1991).

Chapter 12 High-Performance Teams: Getting a Hold on the Process

Griggs, Lewis, et al., *Valuing Diversity: New Tools for a New Reality* (New York: McGraw-Hill, 1995).

Kotter, John P., *Leading Change* (Boston: Harvard Business School Press, 1996).

Pitcher, Patricia, *The Drama of Leadership* (New York: Wiley, 1997).

Wess, Robert, *Leadership Secrets of Attila the Hun* (New York: Warner Books, 1987).

Chapter 13 Visioning: Deciding What the Future Will Hold

Naisbitt, John, & Patricia Aburdene, *Megatrends 2000: New Directions for Tomorrow* (New York: Avon Books, 1990).

Senge, Peter, *The Fifth Discipline: The Art and Practice of the Learning Organization* (New York: Doubleday, 1990).

Senge, Peter, et al., *The Fifth Discipline Fieldbook* (New York: Doubleday, 1994).

Wycoff, Joyce, *Mindmapping* (New York: Berkeley Books, 1991).

Chapter 14 The Assistant Chiefs: Your Spouse and Your Secretary

Covey, Stephen, Roger Merrill, & Rebecca Merrill, *First Things First* (New York: Simon & Schuster, 1994).

Fulghum, Robert, *All I Really Need to Know I Learned in Kindergarten* (New York: Ivy Books, 1988).

Gray, John, *Men Are from Mars, Women Are from Venus* (New York: HarperCollins, 1992).

Hendrix, Harville, *Getting the Love You Want* (New York: Harper & Row, 1990).

Schaef, Anne Wilson, *Women's Reality* (San Francisco: HarperCollins, 1992).

Van Fleet, James K., *Success with People* (Englewood Cliffs, NJ: Prentice-Hall, 1995).

Chapter 15 Are You an Eagle, a Tiger, or a Turtle? How Are You Going to Lead the Organization?

Fink, Stephen, *Crisis Management: Planning for the Inevitable* (New York: Amacom, 1986).

Lucas, James R., *Fatal Illusions* (New York: American Management Association, 1997).

Lustberg, Arch, *Testifying with Impact* (Washington, D.C.: American Society of Association Executives, 1983).

Syles, Leonard, *Leadership: Managing in Real Organizations* (New York: McGraw-Hill, 1989).

Chapter 16 Is This a Marathon or a One Hundred Yard Dash? Reducing Stress

Brahm, Barbara J., *Calm Down: How to Manage Stress at Work* (Glenview, IL: Scott, Foresman, 1990).

Scott, Gini Graham, *Resolving Conflict with Others and Yourself* (Oakland, CA: New Harbinger Publications, 1990).

Selye, Hans, *Stress Without Distress* (New York: Signet Books, 1974).

Sharpe, Robert, & D. Lewis, *Thrive on Stress* (Warner Books, 1977).